Concepts in
Cereal Chemistry

Concepts in
Cereal Chemistry

Finlay MacRitchie

CRC Press
Taylor & Francis Group
Boca Raton   London   New York

CRC Press is an imprint of the
Taylor & Francis Group, an **informa** business

CRC Press
Taylor & Francis Group
6000 Broken Sound Parkway NW, Suite 300
Boca Raton, FL 33487-2742

© 2010 by Taylor and Francis Group, LLC
CRC Press is an imprint of Taylor & Francis Group, an Informa business

International Standard Book Number: 978-1-4398-3582-1 (Paperback)

Library of Congress Cataloging-in-Publication Data

MacRitchie, Finlay.
    Concepts in cereal chemistry / Finlay MacRitchie.
        p. cm.
    Includes bibliographical references and index.
    ISBN 978-1-4398-3582-1 (pbk. : alk. paper)
    1. Cereals as food--Analysis. 2. Grain--Biotechnology. I. Title.

TX557.M337 2010
641.3'31--dc22
                                                   2010011675

Visit the Taylor & Francis Web site at
http://www.taylorandfrancis.com

and the CRC Press Web site at
http://www.crcpress.com

Contents

Preface

This book is based on a course, "Advanced Cereal Chemistry," given to graduate students at Kansas State University in spring semesters since 1998. The aim was not to disperse a lot of information, but rather to present and discuss concepts that provide a foundation for tackling problems that are likely to arise in research on cereals. It should serve as a reference for graduate students in the food and grain science fields as well as for those engaged in research or research/development in these areas. It is hoped that the reader will critically evaluate all that is written and not accept anything without question. In the event of a reader's disagreement, I would welcome receiving comments.

It is rather unfortunate that the term "cereals" has come to be associated with breakfast foods. The justification for this is that breakfast foods are products that are mostly made from cereals. However, products such as breads, cookies (biscuits), pastries, noodles, and pasta are usually not referred to as cereals. The more scientific definition of cereals is that they are cultivated grasses of the Gramineae family that include wheat, rice, maize (corn), barley, sorghum, oats, rye, millet, and triticale. The grain from these cereals provides a large proportion of the world's food.

The book's content has been influenced greatly by interaction with students and colleagues in the cereal area and I am indebted to them. In particular, I wish to acknowledge the invaluable assistance of Shuping Yan and Dr. Xiaorong Wu, who have given generously of their time to assist in preparation of the figures and editing of the text.

The objective has been to relate the topics to a fundamental molecular level. It has therefore drawn on theory from the basic sciences, especially chemistry. Much knowledge about cereals remains to be unraveled by inquiring minds. My hope is that this book may be of help to those who embark on this quest.

Finlay MacRitchie

The Author

Following a career as a research scientist with the Wheat Research Unit (later the Grain Quality Research Laboratory) of the Commonwealth Scientific and Industrial Research Organization (CSIRO) of Australia, **Finlay MacRitchie** was a professor in the Department of Grain Science and Industry at Kansas State University (1997–2009). He is currently professor emeritus in this department. MacRitchie has worked in two main fields: fundamental surface and colloid chemistry and cereal chemistry. He has published more than 150 papers in refereed journals and a textbook, *Chemistry at Interfaces* (Academic Press, 1990). He is listed as an Institute for Scientific Information (ISI) highly cited researcher.

Dr. MacRitchie's awards include the F. B. Guthrie Medal of the Cereal Division of the Royal Australian Chemical Institute and the Thomas Burr Osborne Medal and George W. Scott Blair Memorial Award of the American Association of Cereal Chemists (now AACC International). He has been a member of the editorial boards of *Advances in Colloid and Interface Science, Cereal Chemistry,* and *Journal of Cereal Science.* Presently, he is editor-in-chief of *Journal of Cereal Science.*

The author's research in surface chemistry has focused mainly on the adsorption and behavior of proteins at interfaces. His research in cereal chemistry has focused on the characterization of wheat components and their relationships with end-use quality. Much of this work has involved proteins and their role in determining functionality.

chapter one

Introduction

Cereal grains provide the major food for humans, contributing to more than two-thirds of the world's production of edible dry matter and about half of the world's protein. To aid growers and processors in maintaining this production and increasing it to match the growing population requires the application of many sciences. High-yield cereal varieties with resistance to disease and environmental stresses are the main aims of agronomists, geneticists, plant breeders, and plant pathologists. Cereal chemists are also involved to ensure that the processing properties of the mature grain comply with requirements for the various end-use products that are planned.

However, each of the basic sciences mentioned, as well as others such as entomology and nutrition, may be involved to some extent at every stage from planting through storage and processing. Cereal crops are renewable and their chemical components are biodegradable. As a result, increasing attention is being focused on their utilization for production of biofuels and bioplastics. Cereals for food and nonfood uses are therefore growing industries.

This book is not intended to provide full coverage of all the cereals or to review all the literature in the area of cereal chemistry. This would be an impossible task and would prove detrimental to the coherence of the text. If significant contributions are not mentioned, this does not mean that their importance is not appreciated. The book is meant only to be a link with the knowledge that already exists in books and journals that has been advanced by the efforts of many. Much of the discussion is centered on wheat because most basic research has been carried out on this cereal and some of the most enlightened concepts have emerged from this work. However, these concepts are readily applicable to other cereals. When other cereals are relevant to the discussion, they are introduced. In order partly to rectify what may be perceived as an imbalance, a chapter devoted to nonwheat cereals is included.

The topics are arranged in a fairly logical progression, beginning with the first step of milling. Some of the sciences that are relevant to wheat dry milling are briefly summarized in Chapter 2. The topic of how to measure the efficiency of the dry milling process is discussed in Chapter 3. Grain hardness is the primary property for classifying wheat. Soft-grained wheat is preferred for manufacture of cookies, cakes, and pastries. Medium hard-grained wheat is desired for bread and aerated products in general. Very hard wheat (durum) is utilized for making pasta. Explanations for variation of grain hardness are discussed in Chapter 4.

Once the grain is milled to meal or flour, we are concerned with dough properties. The main components of cereal grains are starch and protein. These are both polymers, so it is logical to apply the knowledge obtained from polymer science to gain understanding of dough behavior. Wheat is unique in that it is the only cereal that yields dough with the viscoelastic properties ideally suited for making aerated products. Chapter 5 deals with the composition and structure of dough and how this determines the performance of a baked aerated product.

In addition to requiring optimum rheological properties, stabilization of gas cells during expansion has to be considered. As a result of their surface-active properties, both proteins and lipids play a role in gas cell stabilization. Gluten protein is the component responsible for dough viscoelasticity. Dough rheology is therefore discussed in Chapter 6 by applying concepts from polymer physics. This gives insight into how gluten proteins behave during application of energy in dough-mixing and how developed dough responds to deformation.

Chapter 7 deals with some aspects of processing. To maintain the focus, the processing of aerated products (breads) will be scrutinized. Factors that determine the final quality will be discussed. Lipids assume special importance in processing in combination with proteins.

Once a product has been made, the ability to maintain its stability comes into question. Thus, Chapter 8 covers the topic of shelf life and focuses on the well-known phenomenon of bread-staling and the factors causing it. Starch is the cereal component believed to play the principal role in the staling of aerated products.

As mentioned before, proteins are the key to explaining dough's physical properties (Chapter 6) and they play a major role in cereal technology. For that reason, particular attention is given to proteins in Chapters 9 and 10. Cereal proteins (especially the gluten type) are known for the difficulty encountered in their solubilization. Chapter 9 deals with the topic of protein solubility. Solubilization is usually a necessary precursor for characterization. Some of the main techniques currently used for characterization of cereal proteins are discussed in Chapter 10.

The relationships between cereal composition and functionality are discussed in Chapter 11. Together with knowledge of the genetic control of cereal components, these relationships lead naturally to the discussion in Chapter 12 of approaches to modifying functionality. Some selected aspects of nonwheat cereals are discussed in Chapter 13. Presently, much emphasis is given to health considerations of diets; Chapter 14 covers some of the health-related topics in discussions of cereals.

Chapter 15 is unusual in that it does not directly refer to cereals. However, the philosophy of science is often neglected in university science courses and it is felt that graduates engaged in research who are likely to follow a career in science need to think about what constitutes the scientific method.

At the end of some of the chapters, exercises are set and a few demonstrations are also suggested to help to illustrate concepts. The exercises do not necessarily have simple answers. They are meant to stimulate thinking and, in some cases, require searching the literature.

At the beginning of the course, a set of questions is posed that is also designed to stimulate thinking. These questions may not have easy answers and are certainly not intended to be a static set. The different questions relate to the topics that are covered in succeeding chapters. The most recent questions that have been posed are the following:

1. What is the physical mechanism involved in variations of grain hardness?
2. Which molecular processes are involved in the viscoelastic properties of wheat flour dough?
3. Why does dough from cereals other than wheat not have viscoelastic properties?
4. How do surfactants enhance or diminish loaf volume and texture in bread-making?
5. Is there any way to eliminate crumb-firming in bread-staling?
6. Why are cereal proteins difficult to solubilize?
7. How can dough strength of wheat varieties be increased?
8. What makes a good scientific hypothesis?

chapter two

Sciences applicable to milling

Introduction

In the processing of grain, milling is often the first step. It is one of the oldest industrial processes and is commonly used to refine barley, rye, maize, oats, millet, rice, sorghum, and wheat. Although this chapter will focus on wheat, the principles of milling are similar for the different grains. No attempt will be made to describe the milling process in detail. After a brief description, some of the different sciences that are relevant to an understanding of the milling process will be introduced.

A cross section of a wheat kernel is shown in Figure 2.1 (Evers, Kelfkens, and McMaster 1998). Dry milling of wheat has two main objectives. A universal objective is to grind the grain to a flour or meal. This may not be required in some cases (e.g., milling of rice). Although whole meals are utilized, they are less popular than flour products, which are more refined. Nevertheless, the trend is toward use of whole-grain products because of their associated health benefits (see Chapter 14). Inclusion of bran and germ components may also compromise the storage properties of the flour (e.g., germ oil is susceptible to rancidity). Therefore, a second objective of the miller is to separate the starchy endosperm from the germ and outer or bran layers to produce white flour.

As a result of the intimate contact between the different constituents together with the awkward geometry of the kernel, it is difficult to obtain a perfect separation. The starchy endosperm comprises from about 80 to 86% by weight of the grain dry weight. However, as the milling extraction (weight of flour/weight of grain) is increased toward this figure, increasing contamination of the flour by the bran and vice versa occurs.

Milling process

After grain is received at a flour mill, it is first cleaned to remove dust, foreign seed, chaff, etc. by methods based on differences in density, size, and shape of the materials, usually on inclined planes and/or by use of air currents. The next step is tempering or conditioning, in which water is added (usually) or removed from the grain to give a distribution of moisture that is optimal for subsequent separation of the constituents. The

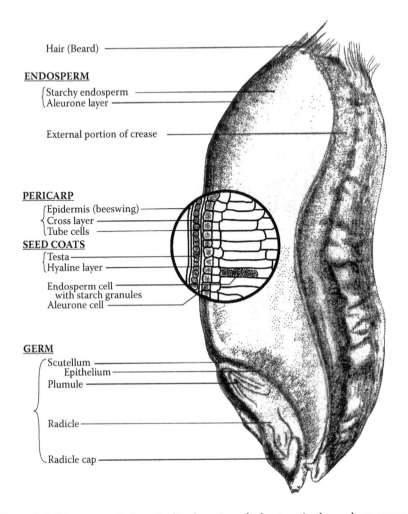

Figure 2.1 Diagrammatic longitudinal section of wheat grain through crease and germ. (Courtesy of Satake Corporation, UK division.)

separation is achieved by continuous processes of grinding and sifting. A roller mill comprises iron rolls working in pairs and geared together so that the upper roll runs at a higher speed than the lower. Two main types of rolls are used:

1. Break rolls are fluted (i.e., they have grooves of saw-toothed sections) and run at a speed differential of 2:1. The spirals on the rolls run in the same direction, leading to a scissor-like shearing action.
2. Reduction rolls are smooth and run at a speed differential of 1.25:1.

The grain is first passed through a series of break rolls, which are progressively more finely fluted and set closer together. In the first break roll, kernels are sheared open and the coarse endosperm is sieved out. The bran with adhering endosperm is passed to the next break roll. The process continues until, usually after four or five break passages, further passages are limited by contamination of the endosperm. The rather coarse endosperm material is removed from as much germ and bran as possible and then passed through a series of reduction rolls to grind the flour to the required fineness. This usually means reducing the particle size to the order of 130 μm for flour.

In milling of semolina, which is preferred for pasta production, the particle size is larger. Smaller-size flour particles are not desirable and are removed as a lower quality by-product.

Tempering

In tempering, a certain amount of water is added to the grain, which is then allowed to lie for some time (often overnight). The amount of water and the lying time are critical for optimum milling. The aim is to create a moisture gradient in the kernel as well as to soften the grain so as to reduce the energy required for grinding. The outer bran layers are made relatively moist so that, upon grinding, they do not powder; this facilitates their removal from the flour. In contrast, the endosperm needs to be relatively dry so that it can be finely ground and separated from the bran and germ by sieving.

The optimum tempering water for maximum flour yield and the rate of penetration of moisture into the grain both vary with the hardness of the grain. Hard wheats are tempered to higher water content than soft wheats. The rate of moisture movement into the kernels has been studied using autoradiography to follow the weak beta-emission from tritiated water (Butcher and Stenvert 1973; Moss 1977). Water moves more rapidly through soft grains than through hard grains and also more rapidly through mealy grains (low protein) than through vitreous grains (high protein). This effect is illustrated in Figure 2.2.

It has also been found that penetration of moisture is faster the higher the initial moisture content of the grain is. These considerations are important in making decisions on lying time (i.e., the time between applying the tempering water and milling the grain). Studies of water movement by autoradiography have shown no large barrier to moisture penetration in the outer bran layer and an initial rapid movement of water into the germ and along the dorsal region of the kernel.

We will now examine some of the sciences that are relevant when we try to understand the complex processes occurring during milling.

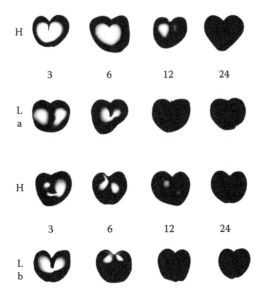

Figure 2.2 Autoradiographs of wheat grains showing the effects of grain hardness and vitreosity on water penetration rates: (a) cultivar Eagle (hard); (b) cultivar Gamenya (soft). H = higher protein level (vitreous grains); L = lower protein level (mealy grains). Numbers indicate lying time in hours. (Data from Moss, R. 1977. *Journal of Food Technology* 12:275–283.)

Particulate-composite systems

Several disciplines are needed to help understand the processes that occur in milling of grain. First, the main chemical components of cereal grains are starch and protein, which are both polymers. The literature on polymer science is vast and much of it can be applied to cereals and specifically to milling. Properties such as hardness, tensile strength, abrasion, and tear resistance depend on the morphology of the kernel and on materials science. What happens to the grain as it is crushed between rollers involves the science of fracture mechanics. Ease of sieving of flour particles is an area that is covered by flow properties of powders.

A useful model for the endosperm structure of a wheat kernel is that of the particulate-composite system (Clyne 1996), in which particles (filler) are dispersed, usually in a polymer matrix. It has its origin in the important industrial process of incorporating tiny particles of carbon black into rubber to enhance toughness and durability. Since then, many polymers have been blended with particles or fibers to modify their properties in favorable ways. Knowledge that has emerged can be applied to endosperm material, which consists of a polymer matrix (storage protein) with a high loading of filler (the dispersed starch granules).

Properties of particle-filled composites, such as strength, toughness, and impact resistance, depend on the material properties of both matrix and filler. Of importance are the size and shape of the filler particles and interactions between them, the total volume occupied by the filler, the presence of voids in the structure, and the degree of adhesion between filler and matrix. When the elastic modulus of the filler is higher than that of the matrix, the modulus of the composite usually increases in proportion to the volume fraction of filler. Opposite behavior is expected if the matrix is more rigid than the filler.

Fillers may be divided into two classes: reinforcing and nonreinforcing. The difference relates to the interfacial adhesion between filler and matrix. A poor matrix–filler interaction results in cavitation and vacuole formation upon significant deformation, thus inducing cracks. Strong bonding between matrix and filler improves reinforcement, leading to enhancement of properties such as elastic modulus. The adhesion between filler particles and matrix is an ongoing area of research and its basis may be different from one system to another. This topic is highly relevant when we discuss variations in grain hardness in Chapter 4.

Fracture mechanics

The strength under stress and the fracture properties of materials make up an extensive branch of engineering science. A vast literature on the subject exists and the reader is encouraged to consult relevant textbooks (e.g., Moore, Pavan, and Williams 2001). Here, only a few brief remarks will be made to serve as an introduction.

When a force is applied to a solid material, the result is deformation. The force is usually expressed as a stress (force per unit area) and the deformation as a strain (change of dimensions per unit dimension). Stress may be compressional (no shape change), tensile (elongation in one dimension), or shear (tangential displacement by parallel forces). In the milling of grain, the stress exerted and the material properties of the grain determine how it fractures. All materials contain flaws. Fracture mechanics involves analysis of these flaws and how stresses cause them to propagate as cracks, leading to catastrophic failure.

At a molecular level, polymeric materials fail by a combination of two mechanisms: bond breakage and chain slippage. When a stress is applied, the initial deformation usually involves shear flow of polymer chains past one another because breaking noncovalent bonds requires much lower stresses than those required to break covalent bonds (by one to two orders of magnitude). The balance between the two mechanisms determines the fracture behavior. If breakage of bonds is the predominant mechanism, brittle fracture is observed. If slippage of chains is the main process, more ductile fracture occurs. Of course, failure usually

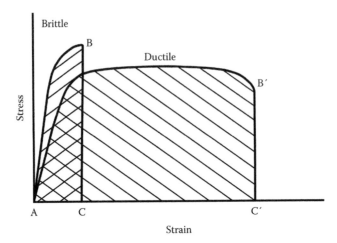

Figure 2.3 Schematic representations of tensile stress–strain behavior for brittle and ductile materials loaded to fracture.

involves a combination of both mechanisms. The difference between brittle and ductile fracture is illustrated in Figure 2.3. The total energy required to fracture the sample is given by the area under the stress–strain curve.

Powder technology

An important property in milling is the ease of sieving particles as the grain passes through the different stages. This depends on the flow properties of powders—a topic that has been extensively studied (Fayed and Otten 1984). The close relationship between the ease of sieving and the hardness of the grain will be considered in the next chapter. Hard-grained wheats tend to produce regularly shaped particles of relatively large size distribution; these pack well and have good flow properties and thus ease of sieving. In contrast, soft-grained wheats have a smaller size distribution with a large number of small particles that may include individual starch granules.

If V_p is the volume of particles and V_b the volume of the powder bed, the degree of packing is often characterized by the packing fraction (V_p/V_b). Another useful parameter is the fractional voidage or porosity defined by $(1 - V_p/V_b)$. An idealized model made up of equally sized cubes would have a packing fraction of one and thus a porosity of zero. The theoretical porosity for closest packed, uniform, spherical particles is 0.26. The porosity of other simple models may be calculated; however, in practice, the heterogeneity of size and shape of particles and the interactions between them usually make theoretical predictions of porosity difficult. Values of porosity between 0.1 and 0.9 may be obtained in practice.

As grain hardness decreases below a certain value, the packing density of the flour drops rapidly, resulting in loss of mobility of particles and poorer sieving properties. In a comparison of particle size distributions, it was found that the proportions of particles of diameter below 25 μm (material in the free starch granule and protein fragment range) were 15.8% for a hard wheat and 35.9% for a soft wheat (N. L. Stenvert, personal communication). Increasing the percentage of particles in the small size range (<25 μm) was found to decrease markedly the packing density and the throughput rates of sieving.

It might be expected that inclusion of small particles in a powder could decrease porosity if they fit into voids between larger particles. In practice, the opposite effect is usually observed. The explanation for this follows from consideration of the main factors that determine closeness of packing and flow properties of powders: the size, shape, and surface properties of the particles. It is often found that the effects of surface properties outweigh the others because they govern the friction and adhesion between particles. As the size of particles decreases, the ratio of surface to volume increases, thus magnifying frictional resistance. Other factors that may contribute to increased friction or stickiness are the presence of liquid films and electrical charge effects.

Several methods are used to assess stickiness in powders (Neumann 1953). Perhaps the most common one is to measure the angle of repose. A conical pile of the powder is formed and the angle between the free surface and the horizontal is measured as shown in Figure 2.4. This angle represents an equilibrium in which gravitational forces are balanced by the frictional forces between the particles. Thus, a high angle reflects a high degree of stickiness.

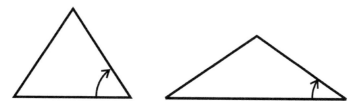

Figure 2.4 High and low angles of repose for powders differing in flow properties.

References

Butcher, J., and N. L. Stenvert. 1973. Conditioning studies on Australian wheats. III. The role of the rate of water penetration into the wheat grain. *Journal of the Science of Food and Agriculture* 24:1077–1084.

Clyne, T. W. 1996. Interfacial effects in particulate, fibrous and layered composite materials. *Key Engineering Materials* 116–117:307–330.

Evers, A. D., M. Kelfkens, and G. McMaster. 1998. Ash determination—A useful standard or a flash in the pan? Satake Corporation, UK division, Presented at ICC Technical Conference, Valencia, Spain (http://www.satake.co.uk/laboratory/Ash%20Tony%20Evers.htm).

Fayed, M. E., and L. Otten. 1984. *Handbook of powder science and technology.* New York: Van Nostrand Reinhold.

Moore, D. R., A. Pavan, and J. G. Williams. 2001. *Fracture mechanics testing methods for polymers: Adhesives and composites.* New York: Elsevier.

Moss, R. 1977. The influence of endosperm structure, protein content and grain moisture on the rate of water penetration into wheat during conditioning. *Journal of Food Technology* 12:275–283.

Neumann, B. S. 1953. Powders. In *Flow properties of dispersed systems,* ed. J. J. Hermans, 382–422. Amsterdam: North-Holland.

chapter three

Efficiency of dry milling

Ash content of flour

A perfect milling process would separate the starchy endosperm comprising approximately 80–86% of the kernel from the bran and germ (see Figure 2.1 in Chapter 2). The efficiency depends on how closely this objective is achieved. Figure 3.1 illustrates the separation process and Figure 3.2 shows how much each botanical fraction contributes to the total dry mass. In order to obtain a measure of the efficiency, the traditional method has been to measure the ash content of the flour. Ash content is determined by combustion of the flour in a high-temperature furnace to vaporize the organic constituents, leaving an inorganic residue: the ash or minerals.

The rationale for using ash values can be understood from Figure 3.3, which depicts the percentage of ash in each botanical fraction. The aleurone layer that surrounds the starchy endosperm has the highest percentage. Other components of the bran and the germ are high, whereas the ash content of starchy endosperm is very low. Therefore, it should be possible, at least to some extent, to monitor the bran and germ contamination of the flour by means of the ash content.

A detailed study by scientists at the Netherlands Organization for Applied Scientific Research (TNO) of a range of wheat cultivars, grown at a number of sites in three consecutive years, raised two objections to the use of ash for measuring efficiency of separation (A. D. Evers, Kelfkens, and McMaster 1998; T. Evers and MacRitchie 2000). First, the ash content of different bran components varied appreciably. As seen from Figure 3.3, the outer pericarp has much lower ash than the other bran layers. This means that it is possible to include a relatively large amount of pericarp in the flour without greatly raising the ash level. This is demonstrated in Figure 3.4, where the amounts of different fractions required to raise the ash level of a flour to 0.5% are compared. Concentration of the outer pericarp (or beeswing) can have deleterious consequences because it has tiny, needle-like hairs that, if present in the flour, cause puncturing of gas cells during bread-making (Gan et al. 1990).

A second problem arose when it was found that a close relationship between flour ash and grain ash existed. There could be two possible reasons for this:

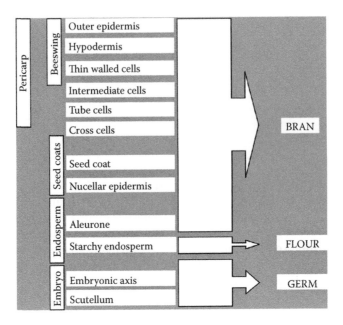

Figure 3.1 Relationships between grain parts and milling fractions for dry milling of wheat. (Courtesy of Satake Corporation, UK division.)

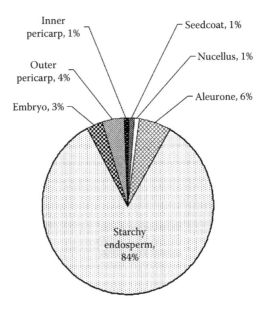

Figure 3.2 Pie chart of contributions to dry mass of grain parts. (Courtesy of Satake Corporation, UK division.)

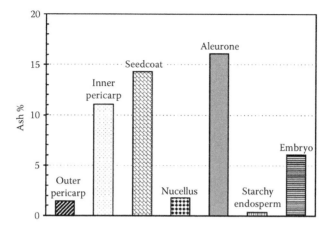

Figure 3.3 Ash values of individual botanical components of wheat grain. (Courtesy of Satake Corporation, UK division.)

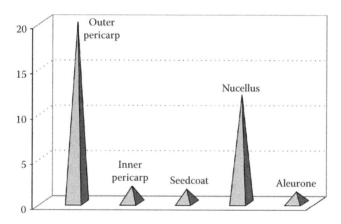

Figure 3.4 Proportions of individual bran tissues that would be present in a flour of 0.5% ash (endosperm ash 0.3%). (Courtesy of Satake Corporation, UK division.)

1. Grain with high ash contains a higher proportion of nonendosperm material.
2. One or more of the components of the high-ash grain had higher ash values.

It was found that the second possibility was correct and that the component that varied in ash was the endosperm. The variation in endosperm ash could be up to 50%. The effect of this is that flours with high endosperm ash may be unfairly penalized if grain is milled to a given flour ash content.

Table 3.1 Correlations between Flour Quality Parameters and the Two Methods of Measuring Milling Efficiency: Ash and Bran by Image Analysis

Quality parameter	Ash	Bran by image analysis
Loaf volume	n.s.	−0.71
Zeleny	n.s.	−0.65
Water absorption	n.s.	−0.68
R_{max}	n.s.	−0.62
Extensograph area	n.s.	−0.54

Source: Reproduced by permission from Evers, A. D. et al. 1998. Presented at ICC Technical Conference, Valencia, Spain (http://www.satake.co.uk/laboratory/Ash%20Tony%20Evers.htm).

Note: n.s. = nonsignficant.

Image analysis

Another method for assessing milling efficiency that has recently been used more is image analysis. The basis for this method is that bran and germ constituents are inherently dark in color, whereas starchy endosperm is pure white. Image analysis instruments magnify a flour sample to give a resolution in pixels. They then compute the number and intensities of dark particles against a white starchy endosperm background. This is a more direct method than ash measurement. It is also simpler, is nondestructive, and can be used in the lab or online in a mill. Table 3.1 shows correlations that were found between some flour quality parameters and the two methods for measuring efficiency of milling separation.

References

Evers, A. D., M. Kelfkens, and G. McMaster. 1998. Ash determination—A useful standard or a flash in the pan? Satake Corporation, UK division. Presented at ICC Technical Conference, Valencia, Spain (http://www.satake.co.uk/laboratory/Ash%20Tony%20Evers.htm).

Evers, T., and F. MacRitchie. 2000. Ash—Who needs it? *World Grain* 18:56–63.

Gan, Z., R. E. Angold, M. R. Williams, P. R. Ellis, J. G. Vaughan, and T. Galliard. 1990. The microstructure and gas retention of bread dough. *Journal of Cereal Science* 12:15–24.

chapter four

Grain hardness

Introduction

Dry milling of grain depends on a property of the grain called "hardness." Hardness determines the forces required to fracture the grain, the manner in which fracture occurs, and the size and shape of the fragments. We will be mainly concerned here with the hardness of wheat in relation to its milling. Grain texture (i.e., "hard" or "soft") is the primary means of classifying wheat for commerce. It is possibly the single most important trait for end-use quality and utilization (Morris et al. 2001). Durum (tetraploid) wheats are very hard and are preferred for pasta-making. Other wheats tend to be referred to as soft in Europe, but in other parts of the world (e.g., North America), they are further classified into hard and soft. Hard wheat varieties are preferred for bread-making; soft wheat varieties are used in the manufacture of cookies, cakes, and pastries.

The main reason for these preferences is that different degrees of hardness give different levels of starch damage when milled. In harder wheats, as we shall see when we consider the origin of hardness, the kernel structure is more coherent. The starch granules are more firmly embedded in the surrounding matrix. This causes high stresses on the starch granules during milling, thus subjecting them to more damage. In softer wheats, adhesion between starch granules and matrix is weaker so that, upon applying stresses, the granules break away from the matrix more easily and experience less damage.

A fermentation process (as in bread-making) requires the presence of moderate amounts of fermentable carbohydrates (simple sugars) for the leavening agent (yeast) to feed on. If they are not present naturally, these sugars can be formed by breakdown of the large starch polymers by enzymes (amylases). The rate of breakdown is greater for damaged than for intact starch. For products requiring less or no fermentation, such as cookies, less starch damage is needed. A certain proportion of wheat is utilized for gluten/starch manufacture. For most starch applications, low starch damage is satisfactory.

Methods for measuring grain hardness

Several different methods are available for measuring grain hardness. These have been classified into four groups depending on whether they are based on grinding, crushing, abrasion, or indentation by a stylus (Simmonds 1974). In the particle size index (PSI) test, a 10-g sample of ground grain is sieved over a standard mesh sieve (AACC method 55-30). The weight of the throughs is used to calculate the PSI (Symes 1961). As mentioned in Chapter 2, soft-grained wheats give a smaller size distribution of particles, so there is an inverse relationship between hardness and PSI. Another test based on grinding that has been used is the Stenvert hardness test (Stenvert 1974; Pomeranz, Czuchajowska, and Lai 1986). Pearling index measures the ease with which the outer layers of the grain are abraded. In this case, the Pearling index correlates positively with hardness.

Near infrared reflectance (NIR) spectroscopy is a convenient method for measuring hardness. It is an indirect method but, as long as a sound calibration is performed against other direct methods, it can provide reliable measures of hardness. The most recent method that has been introduced is that of the single kernel characterization system (SKCS) manufactured by Perten Instruments (Stockholm, Sweden). As its name implies, a sample is fed through the grinder and the characteristics of each individual kernel are measured in a fully automatic mode. For each kernel, several parameters are analyzed: moisture, hardness, kernel diameter, and weight. The great advantage of this instrument for measuring hardness is that, unlike the methods previously mentioned that give a single value, the SKCS system provides a distribution of hardness. This is particularly important because there is always a gradation of hardness between kernels from the same wheat and the distribution profile gives information not given by the other methods.

What is the origin of grain hardness?

Before exploring the physical mechanism behind variations of grain hardness, it is useful to consider its genetic control. It has been found that one major gene controls the hardness. This gene is located on the short arm of chromosome 5D of hexaploid (bread) wheats (Symes 1961). The genetic structure of wheat will be described in Chapter 11. Suffice it to say here that the genetic origin of the hardness trait is simple. There is a single site (locus) at which the major gene is located (i.e., on the short arm of one of the 42 chromosomes of a bread wheat). This should help in trying to understand the origin of differences in hardness.

It is interesting to trace the history of research on the origin of grain hardness because this illustrates how science works to increase understanding of a topic by successive and cumulative inputs from researchers. The initial breakthrough resulted from innovative work by Simmonds and co-workers (Barlow et al. 1973; Simmonds 1974). Wheat was pin-milled to yield fine particles comprising fragments concentrated in starch, including individual starch granules, and particles with a high concentration of protein matrix. The particles were then separated by immersion in a liquid of density intermediately between starch and protein. The higher density starch settled on the bottom and the lighter particles, concentrated in protein, floated to the top. A micropenetrometer was then used to measure the hardness of the particles. It was found that matrix protein was about 20% harder than starch.

However, there were no differences in hardness between starch from hard and soft wheats or between matrices from hard and soft wheats. To rationalize this observation, it was postulated that variations in hardness were due to differences in adhesion between starch granules and protein matrix. In hard wheats, strong adhesion caused a more compact structure; in soft wheats, lower adhesion strength resulted in a more porous structure with flaws in the contact between starch granule and matrix.

These differences have been clearly demonstrated by the work of Stenvert and Kingswood (1977) using scanning electron microscopy. We see that in hard-grained wheats the starch acts as a reinforcing filler, and in soft-grained wheats as a nonreinforcing filler according to the terminology used in the topic of particulate composite systems (see Chapter 2).

Starch granule surface proteins

The attention drawn to the starch granule–matrix interface soon resulted in studies of the compounds that concentrate at this interface. It was found that when starch that was water-washed from flour was treated with a solution of the detergent sodium dodecyl sulfate (SDS), a series of proteins was extracted that could be characterized by SDS polyacrylamide gel electrophoresis (SDS-PAGE). When the resolution of these proteins—termed starch granule proteins (SGPs)—was improved by applying gradient SDS-PAGE, an exciting result emerged: A band appeared at a mobility corresponding to a molecular weight (MW) of about 15,000, as seen in Figure 4.1. This band was found to be strong for soft wheat starches and faint for hard wheat starches.

Figure 4.2 shows some results for an interesting set of recombination lines from a cross between the soft-grained variety Chinese Spring and a Chinese Spring line in which the 5D chromosome of the hard wheat Hope has been substituted. Histograms are plotted as a function of grain hardness measured by PSI. Those above the zero line represent lines with a strong band and those below represent lines with a

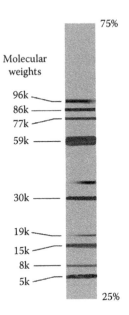

Figure 4.1 Gradient SDS-PAGE of starch granule proteins clearly showing the band for the 15,000-MW protein. (Reproduced with permission from Greenwell, P., and J. D. Schofield. 1986. *FMBRA Bulletin* 4:139–150.)

faint band for the 15,000-MW protein. A clear differentiation of hardness based on the intensity of the band is apparent. It was surmised that the protein of molecular weight 15,000 (later shown to comprise several proteins) must have a role opposite to that of a glue; that is, it must act as a "nonstick" agent, disrupting the adhesion between the starch granule and the protein matrix). The mechanism is illustrated in Figure 4.3.

This nonstick protein was called friabilin (Greenwell and Schofield 1986, 1989). Rahman and co-workers (Jolly et al. 1993; Rahman et al. 1994) isolated friabilin, which they called grain softness protein, and used it to raise polyclonal antibodies. These were used in western analysis to show that the protein that was concentrated at the surface of water-washed starch was present in wholemeal extracts of hard as well as soft wheats. The level in hard wheats was, however, more variable in some hard wheats than among soft wheats. Nevertheless, the observation of Greenwell and Schofield for water-washed starch was found to hold. Another interesting observation was that bound polar lipids followed the same pattern as friabilin; that is, glycolipids and phospholipids were abundant on the surface of water-washed soft wheat starch but scarce on hard wheat starch (Greenblatt, Bettge, and Morris 1995).

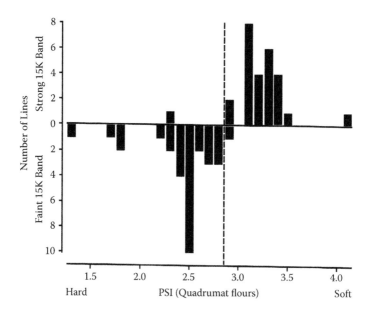

Figure 4.2 Distribution of particle size index (PSI) of flours from 56 Chinese Spring (Hope-5D) chromosome recombination lines. The upper and lower histograms refer, respectively, to lines with and without the strong 15,000-MW band. (Reproduced with permission from Greenwell, P., and J. D. Schofield. 1986. *FMBRA Bulletin* 4:139–150.)

Puroindolines

The next important breakthrough in the evolving research was the identification by Marion and co-workers of two hydrophobic proteins from friabilin characterized by a tryptophan-rich domain (Blochet, Kaboulou, and Marion 1991; Gautier et al. 1994). These proteins were sequenced and named *puroindoline a* and *puroindoline b*. They were to take center stage in understanding the genetic and molecular basis of grain hardness.

A great deal of evidence has accumulated since this early work to confirm that the two puroindolines act together to control the hardness or softness not only of common wheat (*Triticum aestivum*) but also of the seed from other members of the Triticae tribe, including *T. monococcum, Aegilops tauschii,* and various related diploid grasses. It has also become fairly clear that the position of the locus for the major gene for hardness located on the short arm of chromosome 5D corresponds closely to the locus for puroindolines. The reason why durum wheats are always hard becomes obvious now. Durums are tetraploids; that is, they have only A and B genomes and lack the D genome on which the hardness locus is located. Thus, the puroindolines needed to impart grain softness are not expressed in durums.

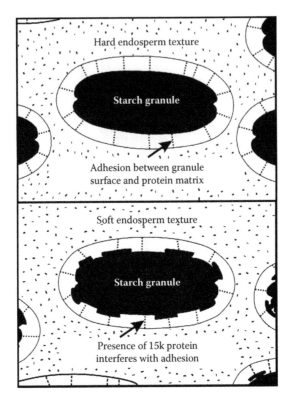

Figure 4.3 Schematic representation of the hypothesis proposed to explain the role of the 15,000-MW protein controlling grain hardness. (Reproduced with permission from Greenwell, P., and J. D. Schofield. 1986. *FMBRA Bulletin* 4:139–150.)

Mutations in puroindolines

The molecular genetic picture that has emerged to describe grain hardness can be summarized as follows. Endosperm softness is the dominantly inherited trait (i.e., the soft phenotype is the wild type). When both puroindoline a and puroindoline b are expressed, they work together to impart softness to the endosperm. When both are absent (as in durums), the endosperm is very hard. When only one is present, an intermediate degree of hardness is found. Another way in which an intermediate degree of hardness is obtained is when one of the puroindolines is present and the other has been altered as a result of mutation.

A number of different hardness alleles have been identified. The gene products of these alleles differ either by the null expression of one of the puroindolines or by a change in a single amino acid in the molecule. One of the earliest mutations to be discovered involved a change in the amino acid at position 46 of puroindoline b from glycine to serine. In an excellent review

Table 4.1 Puroindoline a and b Grain Hardness Alleles, Kernel Phenotype, and Molecular Changes in Expressed Proteins Resulting from Mutations

Puroindoline Locus			
Pina-D1	**Pinb-D1**	Phenotype	Molecular change
Pina-D1a	Pinb-D1a	Soft, wild type	
Pina-D1b	Pinb-D1a	Hard	Puroindoline a null
Pina-D1a	Pinb-D1b	Hard	Puroindoline b, Gly-46 to Ser-46
Pina-D1a	Pinb-D1c	Hard	Puroindoline b, Leu-60 to Pro-60
Pina-D1a	Pinb-D1d	Hard	Puroindoline b, Trp-44 to Arg-44
Pina-D1a	Pinb-D1e	Hard	Puroindoline b null, Trp-39 to stop codon
Pina-D1a	Pinb-D1f	Hard	Puroindoline b null, Trp-44 to stop codon
Pina-D1a	Pinb-D1g	Hard	Puroindoline b null, Cys-56 to stop codon

Source: Data from Morris, C. F. 2002. *Plant Molecular Biology* 48:633–647.

of the topic, Morris (2002) summarized the alleles that had been found up to that time. These are shown in Table 4.1, where the different alleles, the phenotype, and the corresponding molecular change to the puroindoline are given.

It is interesting that a single substitution of an amino acid residue can cause such a dramatic change in endosperm texture. The effect is similar to some of the genetic diseases, of which sickle cell anemia is a well-known example. In this disease, a mutation in the gene coding for the β-globin subunit of hemoglobin produces a change from the amino acid glutamic acid to valine at position 6 of a 146-amino acid chain.

The physical mechanism responsible for differences in endosperm texture (hardness)

The research efforts discussed in the previous sections of this chapter have steadily elucidated the genetic/molecular basis for variations in endosperm hardness. What is still not very well understood at this stage is the physical mechanism that causes soft wheat to become hard as a result of changes in the puroindoline proteins. In order to make progress in this challenging problem, the science of adhesion needs to be invoked.

Theories of adhesion

Although no unifying theory explains all cases of adhesion, several mechanisms have been proposed. These may act completely separately or in combination in certain instances. The main theories of adhesion that are applicable to polymer systems include the following:

Mechanical interlocking occurs when the adhesive penetrates pores and openings in the substrate, producing a "lock and key" effect. This requires the adhesive to "wet" the substrate. Its effect can be enhanced by roughening the surface, thus creating pores and increasing the surface area.

Chemical bonding involves covalent bonding between atoms of the adhesive and substrate.

Weak boundary layers involve contamination of the interface by extraneous materials, disrupting adhesion.

In *adsorption*, intimate intermolecular contact between the two adhering materials entails surface forces that develop between the atoms in the two surfaces.

In *diffusion theory*, the major driving force for adhesion is the mutual diffusion of polymer molecules across the interfaces with resulting interpenetration.

Application to endosperm hardness

Of the different mechanisms, the one that may be most relevant to grain hardness is the adsorption model, with diffusion as a possible secondary factor. During the early stages of kernel development, there is an aqueous environment (water content >50%) in which chemical constituents are relatively mobile. Partitioning of these compounds occurs between the starch granule surface and surrounding matrix. The composition of the adsorbed layer at the starch granule surface will depend on the concentrations of each component and their hydrophobic/hydrophilic balance. Evidence that partitioning may be established early in kernel development comes from a report by Bechtel, Wilson, and Martin (1996) in which immature kernels of hard and soft wheats were air-dried or oven-dried. The difference in hardness of the kernels matched differences in hardness of the mature grain.

In the high water content of the immature kernel, proteins and lipids compete for the starch granule surface. The granule surface would be expected to be less polar (more hydrophobic) than the aqueous medium. Puroindolines are hydrophobic proteins and would therefore tend to adsorb at the starch granule surface. By the mass action law, their amounts would be greatest in the wild phenotype (i.e., with both puroindolines present).

The effect of mutations causing substitution of single amino acids may possibly be rationalized as follows. Protein molecules tend to unfold at an interface to maximize nonpolar residue interaction with the more nonpolar phase and polar residues with the more polar phase. In their unfolded state, polymers (proteins) adsorb as segments (of the order of 6–10 amino acid residues). These segments attach to the surface (trains) or extend into the adjacent phase (loops and tails). The free energy of adsorption per segment may be quite low, but the total free energy per molecule is the sum for all segments and, for large molecules, can attain high values (the concept of free energy is discussed in Chapter 9). However, the free energy of adsorption for a protein is often delicately balanced. Thus, the substitution of a nonpolar amino acid by a more polar one (e.g., glycine to serine) could conceivably alter the sign or at least bring the free energy of adsorption closer to zero.

The other factor to consider is the molecular weight. Puroindolines are rather small proteins. Other proteins in the kernel can reach very high molecular weights (e.g., glutenins) and thus have high free energies of adsorption. They could therefore replace puroindolines at the starch surface.

During desiccation of the kernel, removal of water sets up contracting stresses within it. Whether these stresses lead to discontinuities in the matrix depends on the adhesion between starch granule and matrix. Two main factors influence the strength of adsorption of proteins at the starch granule surface: their hydrophobicity and molecular size. Small molecules would adsorb preferentially as long as they were highly hydrophobic, like the puroindolines in the wild state. The hydrophobicity at the interface would then enhance binding of lipids, but the combination of lipids and small proteins would give no strong interaction between starch and matrix.

When a difference in hydrophobicity is not an issue, large molecules such as the polymeric proteins would have higher free energies of adsorption and would bind strongly to starch. They could also interact strongly with other molecules in the matrix through formation of entanglements (the diffusion theory of adhesion). Whether the stresses that are set up cause the matrix and starch granules to separate in this manner depends on the strength of attachment of the adsorbed protein to the starch as well as the strength of interaction with the matrix. It may be instructive to measure the composition of proteins adhering to starch granules in wheats with a range of hardness. An innovative approach to study the role of the starch–matrix interface has been reported by Malouf and Hoseney (1992), who used reconstitution methods and tensile strength measurements.

Exercises

1. Students are supplied with numbered packets of wheat kernels (6–10 kernels) of known hardness (soft, hard, durum) and asked to classify them as soft, hard, or very hard.
2. Write a paragraph about Table 4.1, attempting to explain why the different mutations induce a change from soft to hard kernels.

References

Barlow, K. K., S. M. Buttrose, D. H. Simmonds, and M. Vesk. 1973. The nature of the starch protein interface in wheat endosperm. *Cereal Chemistry* 50:443–454.

Bechtel, D. B., J. D. Wilson, and C. R. Martin. 1996. Determining endosperm texture of developing hard and soft red winter wheats by different methods using the single-kernel wheat characterization system. *Cereal Chemistry* 73:567–570.

Blochet, J. E., A. J. P. Kaboulou, and D. Marion. 1991. Amphiphilic protein from wheat flour; specific extraction, structure and lipid binding properties. In *Gluten proteins 1990*, ed. W. Bushuk and R. Tkachuk, 314–325. St. Paul, MN: American Association of Cereal Chemists.

Gautier, M.-F., P. Cosson, A. Guirao, D. Marion, and P. Joudier. 1994. *Triticum aestivum* puroindolines, two basic cysteine-rich seed proteins: cDNA analysis and development gene expression. *Plant Molecular Biology* 25:43–57.

Greenblatt, G. A., A. D. Bettge, and C. F. Morris. 1995. The relationship among endosperm texture, friabilin occurrence, and bound polar lipids on wheat starch. *Cereal Chemistry* 72:172–176.

Greenwell, P., and J. D. Schofield. 1986. What makes hard wheats soft? *FMBRA Bulletin* 4:3–18.

———. 1989. The chemical basis of grain hardness and softness. In *Wheat end-use properties*, ed. H. Salovaara, 59–72. Proceedings of ICC '89 Symposium, Lahti, Finland.

Jolly, C. J., S. Rahman, V. Kortt, and T. J. V. Higgins. 1993. Characterization of the wheat M_r 15,000 grain-softness protein and analysis of the relationship between its accumulation in the whole seed and grain softness. *Theoretical and Applied Genetics* 86:589–597.

Malouf, R. B., and R. C. Hoseney. 1992. Wheat hardness. I. A method to measure endosperm strength using tablets made from wheat flour. *Cereal Chemistry* 69:164–168.

Morris, C. F. 2002. Puroindolines: The molecular genetic basis of wheat grain hardness. *Plant Molecular Biology* 48:633–647.

Morris, C. F., G. E. King, R. E. Allan, and M. C. Simione. 2001. Identification and characterization of near-isogenic hard and soft hexploid wheats. *Crop Science* 41:211–217.

Pomeranz, Y., Z. Czuchajowska, and F. S. Lai. 1986. Gross composition of coarse and fine fractions of small corn samples ground on the Stenvert hardness tester. *Cereal Chemistry* 63:22–26.

Rahman, S., C. J. Jolly, J. H. Skerritt, and A. Wallosheck. 1994. Cloning of a wheat 15 kDa grain softness protein (GSP). GSP is a mixture of puroindoline-like polypeptides. *European Journal of Biochemistry* 223:917–925.

Simmonds, D. H. 1974. Chemical basis of hardness and vitreosity in the wheat kernel. *Baker's Digest* 48:16–29, 63.

Stenvert, N. L. 1974. Grinding resistance, a simple measure of wheat hardness. *Flour and Animal Feed Milling* 12:24–27.

Stenvert, N. L., and K. Kingswood. 1977. The influence of the physical structure of the protein matrix on wheat hardness. *Journal of the Science of Food and Agriculture* 28:11–19.

Symes, K. J. 1961. Classification of Australian wheats on the granularity of their whole meal. *Australian Journal of Experimental Agriculture and Animal Husbandry* 1:18–23.

chapter five

Structure and properties of dough

Introduction

Wheat flour dough is the starting material for many products that are attractive to consumers. These include pasta and noodles, cookies, and pastries, as well as aerated products comprising the many different types of bread. The focus here will be on Western-style bread. There are three requirements to form a dough: flour, water, and energy. If one is missing, a dough will not be formed. Energy must be imparted to give dough its unique properties (e.g., by mixing or sheeting through rolls). If we are considering a dough to make aerated products, a fourth ingredient is needed: air:

Flour + water + air + energy → dough (for aerated products)

These four ingredients show a resemblance to the four elements of the medieval philosophers: earth (from which flour is derived), water, air, and fire (corresponding to energy). When energy is imparted to a system comprising a solid (flour), a liquid (water), and a gas (air), it is converted into a material (dough) that has properties of both a liquid and a solid. It has viscosity like a liquid and elasticity like a solid and thus is said to be "viscoelastic"—a property that we will pursue further in Chapter 6.

Of course, other ingredients are usually added in a bread formulation, of which a leavening agent (normally yeast) is the most important. Yeast produces the CO_2 that dissolves in the aqueous phase of dough before diffusing into air (gas) bubbles and causing the dough to rise. The purpose of the yeast is mainly to cause expansion of the dough, but this is not the only way in which dough can be expanded. If it is subjected to low pressure, it will inflate as shown in Figure 5.1 (Gandikota and MacRitchie 2005). If placed in a baking pan, the expanded dough will attain the same shape as a baked loaf. Of course, this method is usually not practical for producing bread.

Let us examine the events that occur when a flour-and-water mixture of optimum water content is mixed.

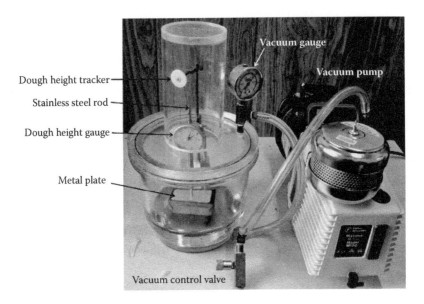

Figure 5.1 Vacuum expansion apparatus for inflating dough. (Gandikota, S., and F. MacRitchie. 2005. *Journal of Cereal Science* 42:157–163.)

Incorporation of water

If increasing amounts of water are added to a flour, mixed, and then spun in an ultracentrifuge, no change is observed until the water content is increased to about 35% (35 g water per 100 g dough). If the centrifugal field is sufficiently high (>100,000 × g) and the dough is spun for sufficient time, a small amount of liquid separates out on top of the dough. As more water is added to the mixed dough, the amount of liquid phase increases until, at about the water content of bread dough, a weight of liquid corresponding to about 20% of the dough weight is separated (Figure 5.2). This is not pure water, but rather a concentrated aqueous solution of soluble flour components (Baker, Parker, and Mize 1946; Mauritzen and Stewart 1965).

Some data for the proportion and composition of the liquid phase are shown in Table 5.1 (MacRitchie 1976). If the time of centrifugation is extended at a given water content, no change in the amount of liquid phase is observed after a certain amount is separated, thus confirming that a true equilibrium has been established.

An insight into the structure of dough may be obtained by measuring its electrical conductivity. Figure 5.3a shows how dough conductivity changes as the dough water content is increased. At low water content (30%), conductivity is very low, but it then begins to rise steeply, extrapolating to a water content of about 35% at zero conductivity. Two conclusions may be deduced from this result:

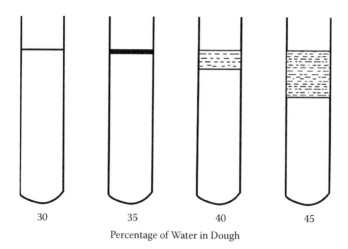

Percentage of Water in Dough

Figure 5.2 Effect of dough water percentage (numbers below tubes) on the separation of the dough liquid phase by ultracentrifugation.

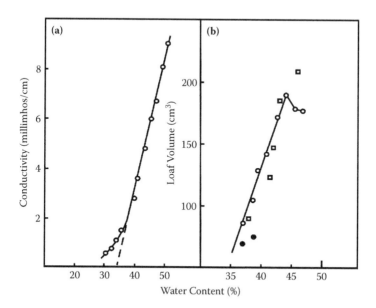

Figure 5.3 (a) Electrical conductivity and (b) test-bake loaf volume of dough as a function of its water content. Open circles: yeasted dough; closed circles: unyeasted dough; open squares: maximum volume of dough pieces expanded by applying low pressure. (MacRitchie, F. 1976. *Cereal Chemistry* 53:318–326.)

Table 5.1 Amounts and Compositions of Dough and Its Solid and Liquid Phases

	Dough		Solid phase		Liquid phase		Amounts not recovered (g)
	Wt (g)	% by wt of dough	Wt (g)	% by wt of solid phase	Wt (g)	% by wt of liquid phase	
Total amount	58.0		47.0		11.0		—
Water	26.7	46.0	16.2	34.5	9.50	86.0	1.0
Protein	4.6	7.9	4.2	8.9	0.37	3.4	0.03
Hydrolysate lipid	0.50	0.9	0.47	1.0	0.03	0.3	—
Sodium chloride	0.70	1.2	0.37	0.8	0.33	3.0	—
Remainder	25.5 (mainly starch)	44 (mainly starch)	25.8 (mainly starch)	55 (mainly starch)	0.77 (mainly soluble carbohydrate)	7.0 (mainly soluble carbohydrate)	

Source: MacRitchie, F. 1976. Cereal Chemistry 53:318–326.

1. A separate conducting aqueous phase separates above the critical water content of about 35%, as already demonstrated by
ultracentrifugation.
2. This aqueous phase must be continuous throughout the dough.
Conductivity is only high if the continuous phase is aqueous.

Thus, when an oil-in-water emulsion inverts to become a water-
in-oil system (e.g., cream to butter), conductivity decreases drastically.
Figure 5.3b also demonstrates the importance of the liquid phase in bread-
making. Dough expansion in a baking test only occurs once a separate
liquid phase is present and the test bake loaf volume, similarly to conductivity, extrapolates to zero expansion near the critical water content of
about 35%.

Incorporation of air

The way in which air (or gas) is incorporated into dough has been ele-
gantly presented in the research of Baker and Mize (1946). Although many
years have passed, these studies still stand as some of the cornerstones of
understanding in baking science. Air is beaten into dough in the form of
tiny bubbles or nuclei during the later stages of mixing, when the dough is
acquiring viscoelastic properties. The occlusion of air is not as obvious as
it is in the beating of a cake batter. Nevertheless, it can be followed, as was
done by Baker and Mize, by simply measuring the dough density.

Figure 5.4 contains much information illustrating this. It shows the
trace (mixogram) registered by a recording dough mixer (the mixograph).
This instrument has pins that measure the resistance to mixing (consis-
tency) with time. The rising part of the trace (dough development stage)
reflects the imparting of viscoelastic properties. After reaching a maximum (peak development time), the trace declines and narrows (breakdown) when the dough progressively loses its elasticity and becomes
sticky. The density trace (plotted to decrease upward) shows that air is
incorporated only in the later stage of dough development. During breakdown, the density tends to stabilize. Dough was taken at different stages of
mixing and baked. Loaf volume and crumb grain improved with increasing development until peak and thereafter declined in quality.

After mixing, no more air is introduced into the dough in bread-making. However, subsequent punching and molding steps can remove some
of the occluded gas and also subdivide the gas cells, thus changing their
size distribution. This is illustrated in Figure 5.5, which shows dough
pieces that have been frozen at the end of proofing and then sectioned.
Dough sample 3 was subjected to an intermediate molding step, whereas
dough sample 4 was not. Improvement of the gas cell structure as a result
of the intermediate molding is easily apparent. When dough sample 3 was

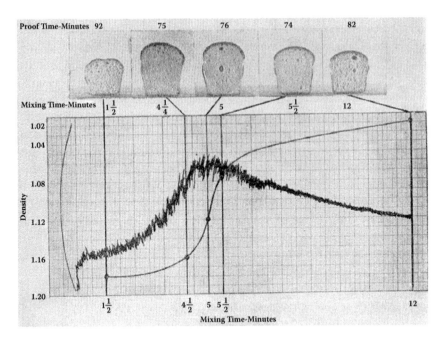

Figure 5.4 Mixogram trace and density curve as a function of time of dough mixing together with loaves baked from dough taken at different times. (Reproduced with permission from Baker, J. C., and M. D. Mize. 1946. *Cereal Chemistry* 23:39–51.)

placed in the oven, it gave good expansion (oven rise) and a high final baked volume. Dough sample 4 gave very little expansion in the oven.

During fermentation, carbon dioxide produced by the yeast dissolves in the liquid dough phase. When solution saturation is reached, the carbon dioxide transfers to the gas cells and causes them to expand. The importance of the presence of gas nuclei can be demonstrated by mixing dough in a vacuum (Baker and Mize 1946). Under these conditions, no gas nuclei are introduced into the dough. Although the yeast can still produce carbon dioxide, there is nowhere for it to be retained and thus it escapes. The result is a baked loaf with low volume and poor crumb grain.

The size of gas bubbles and their concentration in a developed dough are not easy to measure. However, techniques (e.g., tomography) are becoming available that show promise for providing this information. Diameters of gas cells after dough mixing are considered to fall in the range of 10–100 μm. Bloksma (1990) has plotted a relationship between bubble diameter and the percentage of gas volume in a dough for different concentrations of gas cells (shown in Figure 5.6).

In recent times, there has been a rejuvenation in the study of gas bubbles. Two international conferences on "bubbles in food" have been

Figure 5.5 Effect of intermediate molding on gas cell structure. 3: Intermediate molding step included; 4: no intermediate molding step. (MacRitchie, F. 1981. *Cereal Chemistry* 58:156–158.)

organized and many of the contributions have focused on aerated dough products. The physical changes in bread-making are now better understood from the point of view of the creation of a gas cell structure in dough, its manipulation by subsequent operations, and some means of causing expansion.

Changes in dough structure during mixing

Figures 5.7 and 5.8 show micrographs of dough at different stages of mixing. The protein and starch have been stained with specific dyes. Starch granules are stained red and are clearly seen; the gluten protein is stained a blue-green color. In the early stage of dough development (Figure 5.7), the gluten protein appears as discrete clumps of material. As the dough is further mixed, the protein is seen to undergo changes. The clumps become extended and coherent, forming a continuous network throughout the dough (Figure 5.8). The viscoelastic properties of a developed dough are thus the properties of a continuous, hydrated gluten network modified by a high loading of a dispersed phase (or filler)—the starch granules.

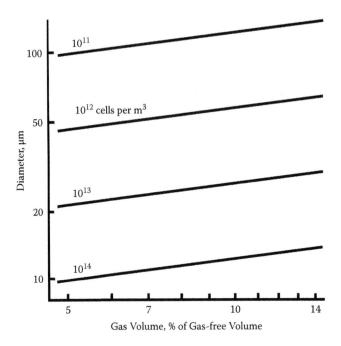

Figure 5.6 Relationship of the total volume fraction of gas, the average diameter of the gas cells, and the number of cells per unit volume of the dough phase. If the first two quantities are known, the third can be estimated. (Reproduced with permission from Bloksma, A. H. 1990. *Cereal Foods World* 35:237–244.)

Model of dough structure

The picture of a developed dough at the end of mixing that has emerged from the preceding studies is roughly depicted in Figure 5.9. A continuous liquid phase envelops the tiny air cells that have been introduced by mixing. These air cells are very highly concentrated, as can be roughly estimated from Figure 5.6. The protein phase is also continuous and, as a result of the directional stress of the mixing, contains protein strands that are aligned. The diagram shows two scales of magnification. On the left, the air bubbles are clearly seen and are of a size estimated to be in the range of 10–100 μm. In the magnified diagram on the right, another feature of the dough structure—the interface between the gas bubble and the liquid phase—is highlighted.

Formation of the air nuclei during mixing depends on the adsorption of surface active compounds present at the gas–liquid interface. The natural surface active compounds in flour are the proteins and lipids. The molecules that adsorb strongly at an air–water interface have a dual nature. They possess a strongly polar (or hydrophilic) moiety, which interacts with the polar phase (water) and a nonpolar (or hydrophobic) moiety

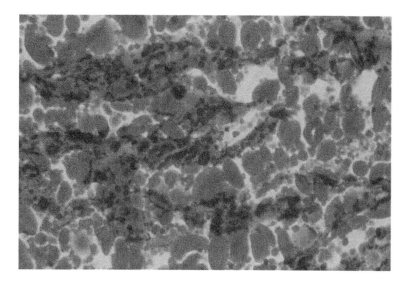

Figure 5.7 **(See color insert following page 82.)** Photomicrograph of dough in the early stage of mixing. Starch has been stained by Ponceau 2R; starch granules are in red dye and protein phase in blue-green. (Courtesy of R. Moss, Bread Research Institute of Australia, North Ryde, NSW, Australia.)

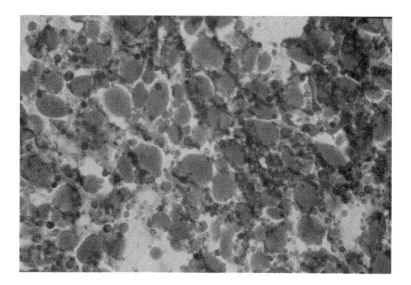

Figure 5.8 **(See color insert following page 82.)** Photomicrograph of dough at the later stage of dough development. Starch has been stained by Ponceau 2R; starch granules are in red dye and protein phase in blue-green. (Courtesy of R. Moss, Bread Research Institute of Australia, North Ryde, NSW, Australia.)

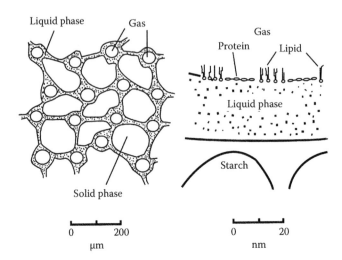

Figure 5.9 Schematic diagram of dough structure at low (left) and high (right) magnification, showing gas cells enveloped by the continuous liquid phase (left) and the stabilizing monolayer at the gas–liquid interface (right). (MacRitchie, F. 1980. In *Advances in cereal science and technology,* ed. Y. Pomeranze, 271–326. St. Paul, MN: American Association of Cereal Chemists.)

that interacts with the nonpolar phase (air). For example, a fatty acid has a polar carboxyl group and a nonpolar hydrocarbon chain. It therefore orients at the interface to present the carboxyl group to the water and the hydrocarbon chain to the air. The concentration and orientation of surfactants at the interface are essential to the formation and subsequent stability of the gas nuclei in the same way that pure water will not form a foam if shaken but readily foams if minute amounts of a surfactant (or detergent) are added. The role of surface active agents in stabilizing gas cells is further discussed in Chapter 7.

References

Baker, J. C., and M. D. Mize. 1946. Gas occlusion during dough mixing. *Cereal Chemistry* 23:39–51.

Baker, J. C., H. K. Parker, and M. D. Mize. 1946. Supercentrifugates from dough. *Cereal Chemistry* 23:16–30.

Bloksma, A. H. 1990. Dough structure, dough rheology and baking quality. *Cereal Foods World* 35:237–244.

Gandikota, S., and F. MacRitchie. 2005. Expansion capacity of doughs: Methodology and applications. *Journal of Cereal Science* 42:157–163.

MacRitchie, F. 1976. The liquid phase of dough and its role in baking. *Cereal Chemistry* 53:318–326.

————. 1980. Physicochemical aspects of some problems in wheat research. In *Advances in cereal science and technology*, ed. Y. Pomeranze, 271–326. St. Paul, MN: American Association of Cereal Chemists.

————. 1981. Flour lipids: Theoretical aspects and functional properties. *Cereal Chemistry* 58:156–158.

Mauritzen, C. M., and P. R. Stewart. 1965. The ultracentrifugation of doughs made from wheat flour. *Australian Journal of Biological Science* 18:173–189.

chapter six

Dough rheology

Introduction

In this chapter, we will attempt to explain dough physical properties and their variation at a fundamental level. In order to do this, it is preferable to take the most general approach. Many materials have properties that show similarity to those of dough; for example, many polymers have viscoelastic properties. Much research on the physical behavior of polymers has been carried out as a result of the great importance of these compounds in the huge plastics industry. The knowledge gained from polymer studies can therefore be directly applied to understand the behavior of dough. Some molecular properties—segmental motion, molecular weight distribution, entanglements, and glass transition temperature—are unique to polymers and it is helpful to keep these in mind as we consider dough properties.

Segmental motion

Because they consist of long chains, polymers (or large molecules in general) show some properties that are dependent on segments of molecules in addition to whole molecules. For example, the energy of activation for flow is found to depend on portions or segments of molecules (Kauzmann and Eyring 1940). This is a reflection that these segments can move, to some extent, independently as units. For proteins, the segment size has been estimated to be in the order of 6–10 amino acid residues (MacRitchie 1998).

Molecular weight distribution

Synthetic polymers usually do not have a discrete molecular weight (MW); they are mixtures of differently sized polymers having a distribution of molecular weights. On the other hand, some biological molecules, such as many proteins, have definite molecular weights because they are synthesized from genes. Gluten proteins contain a heterogeneous mixture of proteins comprising relatively low molecular weight gliadins and very high molecular weight glutenins. Furthermore, the gene products of the glutenin proteins of wheat are subunits. These subunits polymerize by a posttranslational process, producing glutenin polymers with a range of sizes.

Molecular weight distribution (MWD) is often characterized by average molecular weights. The two most commonly used are the number-average (M_n) and the weight-average (M_w) molecular weights. M_n is simply given by the weight of the whole divided by the number of molecules. This is the average that is obtained from measurements of colligative properties such as osmotic pressure or freezing point depression. M_w gives an extra weighting to the larger molecules. This is the average obtained from any property whose intensity is proportional not only to the amount of polymer present but also to the mass of the particle. Thus, properties such as sedimentation and light scattering give a measure of M_w. Another average, the Z-average, gives a still higher weighting to the larger molecules and is also sometimes used.

Number average: $M_n = \Sigma_i N_i M_i / \Sigma_i N_i$

Weight average: $M_w = \Sigma_i N_i M_i^2 / \Sigma_i N_i M_i$

Z-average: $M_z = \Sigma_i N_i M_i^3 / \Sigma_i N_i M_i^2$

where N_i and M_i are the number of molecules and molecular weight, respectively, of the ith component.

Entanglements

A physical property such as viscosity increases as the molecular size of the compound increases. This is simply because the frictional force that a molecule encounters in order to flow increases as its surface area increases. The general relationship between viscosity and molecular weight for a polymer is shown as a log–log plot in Figure 6.1. Viscosity increases with increasing molecular weight as expected. However, at a critical molecular weight, which is different for each polymer, the slope of this relationship dramatically increases. This has been interpreted to mean that an additional resistance to flow associated with widely spaced points along the polymer chain operates. This phenomenon is usually referred to as "entanglements."

If there are two or more entanglement points per molecule, a type of network is set up. In order to flow, a molecule needs to pull neighboring molecules with which it is entangled. These molecules, in turn, may be coupled with others and so on throughout the system. When a stress is imposed on such a system, there is a resistance to flow so that the system resembles a cross-linked polymer. However, it is different from a cross-linked system in that, if the stress is maintained, slippage of the chains

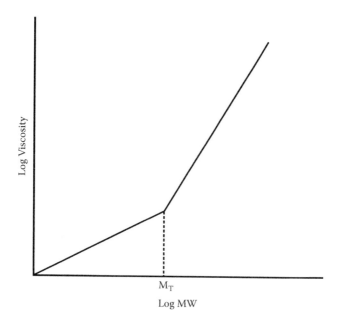

Figure 6.1 Log viscosity versus log MW for a polymer showing the critical MW for effective entanglements.

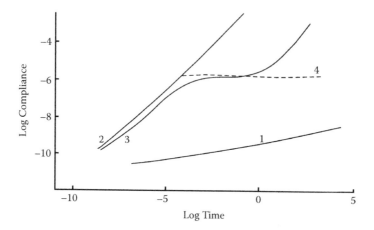

Figure 6.2 Log compliance versus log time for creep. (1) Polymer at a temperature below its T_g. (2) Polymer of low MW. (3) Polymer of high MW. (4) Cross-linked polymer. (Singh, H., and F. MacRitchie. 2001. *Journal of Cereal Science* 33:231–243.)

may occur. This is illustrated in the log–log plot of compliance (strain/ stress) versus time for creep of various polymers in Figure 6.2.

Creep is a rheological method that measures strain versus time for a constant applied stress. Curve 1 in Figure 6.2 is for a polymer below its glass transition temperature, a topic that will be discussed in the next section. The low MW polymer (curve 2) flows relatively easily. The high MW polymer (curve 3), when subjected to a stress, undergoes strain for a time until chains between entanglements are fully stretched, after which a plateau is reached. On continuing application of the stress, slippage of entanglements can occur and the polymer undergoes further strain. This is in contrast to the cross-linked polymer (curve 4), which undergoes no further strain after chains between cross-links have been fully stretched.

An alternative interpretation of the entanglement phenomenon is to consider that a large MW polymer flows by a series of snake-like motions called reptation (De Gennes 1971). The concept of entanglements will be adopted in this text because it has been commonly used in polymer science and may be more intuitively understood by cereal chemists who are new to the area.

Glass transition temperature

Although not unique to polymers, the glass transition temperature (T_g) is particularly important to understanding their physical properties. As the temperature decreases, a point is reached at which segmental motion of the polymer backbone practically ceases. The polymer behaves as a glass. In order for a polymer to exhibit viscoelastic properties, it must be at a higher temperature than its T_g (see Figure 6.3). If a polymer is below its T_g, viscoelasticity can be achieved either by raising the temperature or by increasing the proportion of plasticizer in the system (water, in this case), as shown in Figure 6.3.

As we saw in the previous chapter, protein forms a continuous phase in dough and the properties of dough are essentially those of hydrated gluten protein modified by starch granules acting as a filler. A large number of instruments have been developed to measure physical dough properties in cereal laboratories; the most common ones can be classified into two main groups:

1. recording dough mixers (e.g., farinograph, mixograph)
2. load/deformation instruments (e.g., extensograph, alveograph)

Recording dough mixers measure the resistance to mixing as a function of time. An example was shown in Figure 5.4 in the previous chapter. Characteristically, a dough mixing trace shows first a development stage in which resistance to mixing increases as dough develops its viscoelastic

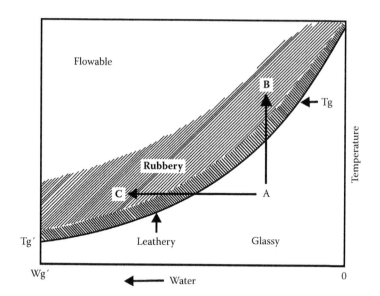

Figure 6.3 An idealized diagram of a cereal material as a function of temperature and water content. T_g is the glass transition temperature at zero water content and T_g' is that at a water content at which free (freezable) water starts to appear as a separate phase (W_g'). (Reproduced with permission from Hoseney, C. R. 1994. *Principles of cereal science and technology,* 2nd ed., 307–319. St. Paul, MN: American Association of Cereal Chemists Inc.)

properties. After a maximum (peak development time) is reached, there is a steady decline (the breakdown stage) in which the dough loses its elastic strength and becomes progressively more sticky.

Figure 6.4 shows some mixograms that illustrate the variation that can occur between the mixing properties of doughs from different flours. Flours with weak dough properties show a steeply rising development stage and a steep breakdown. This is typified by a soft wheat flour suitable for cookie manufacture (shown in Figure 6.4D). In general, as the strength of a flour increases, the development stage becomes longer and the breakdown less steep. A strong flour that could be used for bread-making gives a mixogram similar to that shown in Figure 6.4C. Such a dough is said to have good stability (or tolerance) to overmixing. This means that, in a bakery situation, mixing past the peak will not cause appreciable detrimental effects to the dough properties as would be the case for a weak dough with low mixing stability.

Fundamentals of dough mixing

How do we interpret dough mixing characteristics such as those displayed by mixograms or farinograms? Several observations give us clues and we

Figure 6.4 Mixogram traces. The sequence from A through D may be used to represent either doughs of decreasing mixing requirements at a fixed intensity or one dough mixed at increasing intensities.

will pursue these to try to come up with a molecular explanation of what is involved in dough mixing. A mixture of gliadin and starch with water shows a negligible development stage when mixed, followed by a steep breakdown similar to but more marked than the trace of Figure 6.4D. In contrast, the other component of gluten protein (glutenin) in admixture with starch and water gives a trace similar to that shown in Figure 6.4A, showing no development stage (at that mixing intensity) and no breakdown. Some seminal research on dough mixing by Tipples and Kilborn in the 1970s has enhanced our knowledge of the topic. Two outcomes from this work—critical mixing intensity and "unmixing"—are summarized next.

Critical mixing intensity

In order to develop a dough that has properties suitable for bread-making, two requirements need to be met (Kilborn and Tipples 1972). First, a critical amount of work must be imparted to the dough. Second, optimum development will only be reached if the mixing intensity (i.e., speed) is above a certain critical value. This critical intensity varies with the particular flour.

The mixograms of Figure 6.4 can be used in two ways. First, they can depict the effect of a constant mixing intensity on flours with a range of strengths. The mixing intensity for flour A is below the critical intensity required to develop the dough. Flour B has a lag period but eventually develops—showing that the mixing intensity is close to the critical value. Flour C has a critical mixing intensity below that used and, for a weak flour such as flour D, the critical mixing intensity is well below the intensity used.

The other way in which the mixing curves in Figure 6.4 can be used is to consider the effect of varying the mixing intensity on a single flour. When the mixing intensity is low, it is insufficient to develop the dough (Figure 6.4A). As the mixing intensity is incrementally increased, the mixograms first pass through a stage where the mixing intensity is close to the critical value for the flour (Figure 6.4B). Figures 6.4C and 6.4D are mixograms where the mixing intensity is well above the critical value.

Unmixing

Tipples and Kilborn (1975) demonstrated the phenomenon of unmixing. When a dough is developed by mixing at an intensity above its critical mixing intensity and then subjected to a slower mixing below its critical value, the dough reverts to a state similar to that of a dough that has not been developed. If the mixing speed is then increased above the critical intensity, the dough again develops and the mixing/unmixing cycle can be repeated a number of times. As unmixing proceeds, bread baked from the dough progressively deteriorates in quality.

Molecular aspects of dough development

Based on the preceding discussion, dough development appears to involve the extension of large molecules by shearing and tensile stresses exerted by the mixer. These changes at a molecular level induce the coherent gluten network observed at the macroscopic level. Studies of polymer solutions under stress have shown that a flexible chain should respond to an increase in strain rate through an extension from an almost random state to practically full extension when a critical strain rate, ε_c, is reached (Keller and Odell 1985). The initial resistance to extension is characterized by the conformational relaxation time (τ) appropriate to the random coil given by

$$\tau = 1/\varepsilon_c \qquad (6.1)$$

It has been found that τ is directly proportional to $M^{1.5}$, where M is the molecular weight. It should be noted that we are here extrapolating from solution to a concentrated dough system. However, the general behavior is thought to be analogous.

The increase in resistance during dough development corresponds to the progressive elongation of large glutenin molecules beginning with those of lowest molecular weight. When a series of flours of varying strength are compared, the critical strain rate (mixing intensity) increases as the MWD of the glutenin shifts to higher molecular weights. Thus, the time for a dough to develop (peak mixing time) is an approximate measure of the MWD of its protein. When the MWD is shifted to lower values, as can be achieved by inclusion of cysteine in the dough formulation, the time to development is decreased (Kilborn and Tipples 1973). The phenomenon of unmixing can be understood by considering that the mixing speed is below the critical intensity and is insufficient to maintain the extension of the glutenin molecules. The mixing action allows the molecules to retract and the dough reverts to its unmixed state.

New techniques are being developed that are able to measure the force–distance relationships of single molecules as they are stretched. Such techniques include atomic force microscopy and optical tweezers. Atomic force microscopy utilizes a microscale cantilever that has a probe to scan the specimen surface. Force is measured in piconewtons and the distance in nanometers. An example of some results from this technique is shown in Figure 6.5, where the force required to stretch poly(ethyleneglycol) molecules of different lengths is shown (Oesterhelt, Rief, and Gaub 1999).

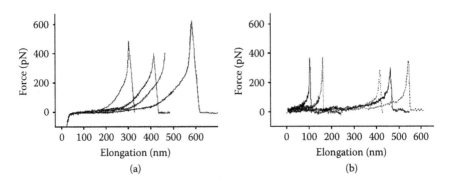

Figure 6.5 Extension traces of individual poly(ethyleneglycol) molecules of different lengths in (a) phosphate buffer solution and (b) hexane. (Oesterhelt, F. et al. 1999. *New Journal of Physics* 6:1–11).

Load-deformation properties

The properties of a developed dough may vary considerably from one flour to another and are measured in cereal labs by load-deformation instruments such as the extensograph and alveograph. These instruments mainly measure two properties: the resistance to stretching and the extensibility. Aerated products (breads) require dough with a desirable balance between these two properties. A dough needs to have extensibility to allow it to expand during fermentation, but it also must have a certain strength to prevent it from collapsing. Although both these properties are measured by the same instrument, they appear to be governed to a certain extent by different molecular mechanisms.

Dough resistance to stretching

The tests used in cereal labs to measure dough properties are similar to those used in industry to test polymers. For example, the extensograph essentially carries out a tensile stress test. An extensograph measurement differs in that it measures force rather than stress and a constant stretching rate is used instead of a constant strain rate. A tensile stress–strain curve is shown in Figure 6.6. In this test, a sample is strained at a constant rate and the stress needed to maintain this strain rate is measured until the point at which it breaks. The peak stress is the ultimate tensile strength,

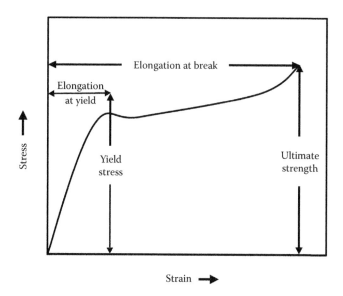

Figure 6.6 Tensile stress relationships for a polymer.

and the distance the sample is stretched before failure is the elongation to break or the draw ratio (length at break/original length).

An equation to describe the tensile strength of polymers was introduced by Flory (1945):

$$\sigma = \sigma_0 \left(1 - M_T/M_n\right) \tag{6.2}$$

where
  σ = tensile strength
  σ_0 = limiting tensile strength (the value reached at very high M_n)
  M_n = number-average molecular weight of the polymer
  M_T = a threshold molecular weight

This equation has been found to hold well for polymers with a narrow range of MWs. However, for polymers with a high degree of polydispersity, the relationship breaks down. Bersted and Anderson (1990) have proposed a modified Flory equation that predicts the tensile strength of polydispersed polymers quite well:

$$\sigma = \sigma_0 \left(1 - M_T/M_n^*\right) \Phi \tag{6.3}$$

The assumption in deriving this equation is that only polymers with MWs above those for effective entanglements contribute to tensile strength. Thus, Φ is the fraction of polymer molecules with MWs greater than M_T, the MW for effective entanglements, and M_n^* is the number-average MW of this fraction. The equation is represented schematically in Figure 6.7, in which Φ is the shaded area.

Dough extensibility

In order to explain the extensibility of polymers, Termonia and Smith (1987, 1988) have taken a different approach. They treat extension in terms of two kinetic processes, each of which is governed by its activation energy. (The important concept of activation energy is described in Chapter 8 [see Figure 8.5]). First, secondary valence bonds must be broken for molecular chains to move relative to one another. Second, once noncovalent bonds between entanglements are broken and chains between entanglements are fully stretched, the only way for further movement of chains relative to one another is by slippage through entanglements. A simplistic picture of the Termonia–Smith model considering only a single molecular chain that is entangled with neighboring chains is illustrated in Figure 6.8.

Before application of stress, chain strands are in a relaxed state (Figure 6.8a). On application of a tensile stress, noncovalent bonds are broken and the chains are extended. The extensional behavior is then

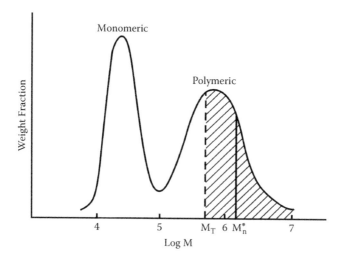

Figure 6.7 Schematic representation of the MWD of gluten protein. M_T is the threshold MW for entanglements; M_n^* is the number-average MW of protein with MW > M_T. Shaded area is Φ, the fraction of MW > M_T.

determined by the relative rates of sample elongation and slippage through entanglements. If the rate of chain slippage is much greater than the rate of elongation, chain slippage occurs rapidly, providing no points of resistance. Tensile strength and draw ratio (length of sample at break/original length) will both be low (Figure 6.8b).

When the rate of chain slippage is low compared to the rate of elongation, chains will not be able to slip through entanglements sufficiently rapidly in response to the applied stress and will break. A high strength may be attained but the draw ratio will be very low (Figure 6.8c). When the rate of slippage is optimum, the draw ratio will be maximized; that is, entanglement points will contribute resistance but chains will slip free sufficiently rapidly to prevent breakage of covalent bonds, giving a high draw ratio with moderate tensile strength (Figure 6.8d).

Application to dough

The theory of molecular entanglements predicts a critical molecular weight for gluten proteins above which they contribute to dough strength. Two experimental results are consistent with this concept. The first, in which 74 recombinant inbred wheat lines were studied, showed a very low correlation ($r^2 = 0.18$) between the proportion of polymeric protein in the flours measured by size-exclusion high-performance liquid chromatography (SE-HPLC) and R_{max}, a measure of dough strength. However, a very high correlation ($r^2 = 0.86$) was found between the proportion

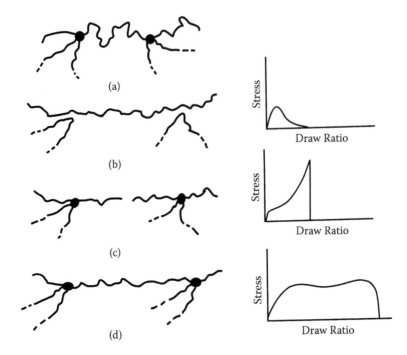

Figure 6.8 Schematic illustration of the effect on the draw ratio of the relative rates of elongation and chain slippage. (a) Initial state; (b) slippage rate >> elongation rate; (c) elongation rate >> slippage rate; (d) slippage rate optimum for elongation rate. (Singh, H., and F. MacRitchie. 2001. *Journal of Cereal Science* 33:231–243.)

of unextractable polymeric protein (UPP) and R_{max} (Gupta, Khan, and MacRitchie 1993). UPP is a measure of the most insoluble and therefore the highest MW fraction of polymeric protein. This result indicated that not all the polymeric protein contributes to dough strength, but rather only a fraction of the highest MW. This is consistent with the concept of the Bersted–Anderson theory.

In a second result, the SE-HPLC chromatograms were obtained for flours from a set of 150 wheat lines (30 varieties grown at five sites). The cumulative chromatogram areas of protein extracts up to different elution times were integrated and correlated with R_{max} (Bangur et al. 1997). As the elution time increased, the correlation coefficient increased until a time corresponding to an MW estimated to be about 250,000 and then decreased thereafter. That is, the maximum correlation with R_{max} was for the fraction of polymeric protein with MW above 250,000. Both these independent experiments therefore gave confirmation of a critical MW of glutenin and that glutenins with higher MW than this value (approximately 60% of the total glutenin) contribute to R_{max}.

Demonstration

The science of rheology is based on measurements of stress (force/area), strain (change in dimensions/original dimensions), and time. Rheological principles can be demonstrated in a simple way by use of "silly putty"—a silicone polymeric material that can be purchased at toy or magic shops. When it is formed into a ball, it will bounce, thus showing that, for a stress applied over a short time, it behaves as an elastic solid. When formed into a cylinder and slowly stretched, it shows plastic deformation, behaving as a viscous material. If the cylinder is stretched rapidly, it will tend to snap, again illustrating behavior of a solid. The difference between purely elastic and viscoelastic properties can be shown by comparing an elastic band and a cylinder of wheat flour dough. Using a ruler, the elastic band can be stretched and, on release of the stress, it will return to its original length. A dough will recover only part of its change in length.

Exercise

1. Water, wheat flour dough, and vulcanized rubber are examples of purely viscous, viscoelastic, and purely elastic materials, respectively. How could the differences in behavior of these three substances be explained at a fundamental level but in simple terms?

References

Bangur, R., I. L. Batey, E. McKenzie, and F. MacRitchie. 1997. Dependence of extensograph parameters on wheat protein composition measured by SE-HPLC. *Journal of Cereal Science* 25:237–241.

Bersted, B. H., and T. G. Anderson. 1990. Influence of molecular weight and molecular weight distribution on the tensile properties of amorphous polymers. *Journal of Applied Polymer Science* 39:499–514.

De Gennes, P. G. 1971. Reptation of a polymer chain in the presence of fixed obstacles. *Journal of Chemical Physics* 55:572–579.

Flory, P. J. 1945. Tensile strength in relation to molecular weight of high polymers. *Journal of the American Chemical Society* 67:2048–2050.

Gupta, R. B., K. Khan, and F. MacRitchie. 1993. Biochemical basis of flour properties in bread wheats. I. Effect of variation in quantity and size distribution of polymeric proteins. *Journal of Cereal Science* 18:23–44.

Hoseney, C. R. 1994. *Principles of cereal science and technology*, 2nd ed., 307–319. St. Paul, MN: American Association of Cereal Chemists Inc.

Kauzmann, W. J., and H. Eyring. 1940. The viscous flow of large molecules. *Journal of the American Chemical Society* 62:3113–3125.

Keller, A., and J. A. Odell. 1985. The extensibility of macromolecules in solution: A new focus for macromolecular science. *Colloid Polymer Science* 262:181–201.

Kilborn, R. H., and K. H. Tipples. 1972. Factors affecting mechanical dough development. I. Effect of mixing intensity and work input. *Cereal Chemistry* 49:34–47.

———. 1973. Factors affecting mechanical dough development. IV. Effect of cysteine. *Cereal Chemistry* 50:70–86.

MacRitchie, F. 1998. Reversibility of protein adsorption. In *Proteins at liquid interfaces,* ed. D. Moebius and R. Miller, 149–177. Amsterdam: Elsevier Science B.V.

Oesterhelt, F., M. Rief, and H. E. Gaub. 1999. Single molecule force spectroscopy by AFM indicates helical structure of poly(ethylene-glycol) in water. *New Journal of Physics* 6:1–11.

Singh, H., and F. MacRitchie. 2001. Application of polymer science to properties of gluten. *Journal of Cereal Science* 33:231–243.

Termonia, Y., and P. Smith. 1987. Kinetic model for tensile deformation of polymers. I. Effect of molecular weight. *Macromolecules* 20:835–838.

———. 1988. Kinetic model for deformation of polymers. II. Effect of entanglement spacing. *Macromolecules* 21:2184–2189.

Tipples, K. H., and R. H. Kilborn. 1975. "Unmixing." The disorientation of developed bread doughs by slow-speed mixing. *Cereal Chemistry* 52:248–262.

chapter seven

Aspects of processing

Introduction

A great variety of products are made from cereals. However, to keep the topic focused, we will concentrate on only one type of product: aerated or leavened breads. Bread from wheat has been studied more extensively than other wheat products or the processing of other cereals. Much of the basic science developed in this area can, however, serve as a basis for other areas of cereal processing.

In Chapter 5 we saw that gas cells beaten into dough in the later stage of mixing formed a structure that enabled the dough to expand during fermentation. The gas cell size distribution and stability are determined by the surface-active compounds present in the dough (proteins and lipids). They adsorb at the gas–aqueous interface and stabilize the gas cells in much the same way as bubbles in a classical foam are stabilized. In all stages of fermentation, expansion of the dough depends on suitable rheological properties of the gluten–starch matrix surrounding the gas cells. In the early stages, the gas cells are small and separated from each other, as shown in Figure 5.9 in Chapter 5. In the later stage of proofing or early stage of baking, gas cells are in close proximity and are surrounded by thin layers (lamellae) of the continuous liquid phase.

The model introduced by Gan et al. (1990) shown in Figure 7.1 applies at this time. Primary and secondary stabilizing mechanisms operate. The primary film is the gluten–starch matrix. When the gas bubbles come into close contact, Gan and colleagues postulated, based on electron microscope pictures, that discontinuities may occur in the gluten–starch matrix surrounding them. Nevertheless, the bubbles may continue to expand because thin liquid lamellae remain around the gas cells. The liquid film provides the secondary mechanism preventing destabilization.

Stability of gas bubbles in dough

Destabilization of gas bubbles may occur as a result of two different mechanisms. The first is disproportionation or Ostwald ripening. Pressure inside a small gas bubble is greater than that outside due to the effect

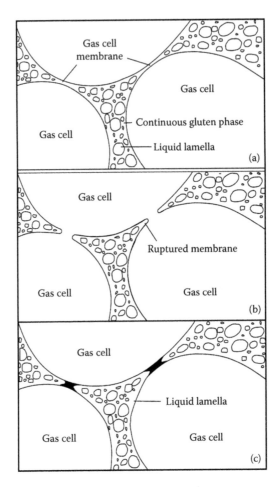

Figure 7.1 Schematic representation of the structure of dough at proof stage. The gluten–starch matrix surrounding gas bubbles (a) becomes discontinuous in the later stages of proof (b), but gas cells remain stable due to the liquid lamellae that envelop them (c). (Reproduced with permission from Gan, Z. et al. 1990. *Journal of Cereal Science* 12:15–24.)

of curvature. The excess pressure inside the gas bubble, known as the Laplace pressure (ΔP), is given by the following equation:

$$\Delta P = 2\,\gamma/r \qquad (7.1)$$

where γ is the surface tension and r is the radius of the gas bubble.

When bubbles of different sizes come in contact, gas will tend to transfer from the smaller bubble (at higher pressure) to the larger, resulting in growth of larger bubbles at the expense of smaller ones. As a consequence, the gas cell size distribution shifts to larger size. The second destabilizing mechanism is coalescence, in which the film between two bubbles coming in contact is ruptured, thus producing a single bubble. This tends also to produce a coarser gas cell structure.

Changes occurring in gas cell structure in dough expansion

We have seen that dough development involves formation of a continuous gluten network (see Figure 5.8 in Chapter 5) and the incorporation of tiny air bubbles that act as nuclei for trapping the gas during fermentation. It is useful to consider these gas bubbles and how they change as the dough expands.

It has been estimated that the occluded gas cells after dough-mixing have diameters in the range of 10–100 μm (Bloksma 1990). The total volume of occluded gas (air) after mixing can be calculated from the density of the dough compared to the density of dough mixed in a vacuum. For many mixers, this is in the order of 10% of the total volume. From these values, it can be estimated that the concentration of gas cells is 10^2–10^5 per cubic millimeter. As fermentation proceeds, gas cells grow and the distance between them becomes smaller. If the gas cells are spherical and of uniform size, they will touch at a relative volume of 3.85. (Relative volume is here defined as the total volume of a dough divided by the volume devoid of gas.)

Figure 7.2 shows how gas cells approach as the relative volume increases. At the end of fermentation, the relative volume is between 4 and 5 and, at the end of oven rise, between 5 and 7. If the gas cells are not uniform in size, they may reach a higher relative volume than 3.85 before coming into contact, but not by very much. It is obvious that, during fermentation, for the dough to continue to expand, the gas cells must distort similarly to the way in which gas bubbles take a polyhedral form in an ordinary liquid foam. During baking in the oven, the gas cells rupture and the dough changes from a foam (gas phase dispersed as bubbles) to a sponge (continuous gas phase).

Primary stabilization of gas cells

The protein–starch matrix provides the primary mechanism for stabilization of gas cells in an aerated product. Rheological properties of dough have been studied extensively with the aim of understanding the properties required to optimize dough expansion during fermentation and baking. Dobraszczyk (1997) has emphasized the need, when carrying out these rheological measurements, to try to mimic the conditions occurring in dough

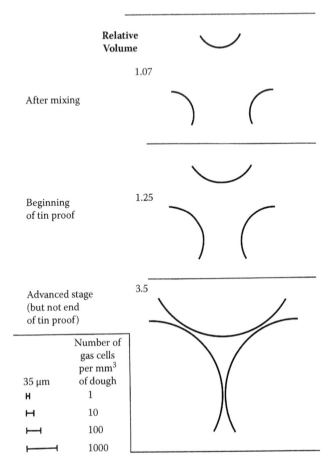

Relative
Volume

1.07

After mixing

Beginning
of tin proof   1.25

Advanced stage   3.5
(but not end
of tin proof)

35 μm	Number of gas cells per mm³ of dough
⊦	1
⊢⊣	10
⊢⊣	100
⊢——⊣	1000

Figure 7.2 Illustration of how the relative volume of dough and the concentration of gas cells change during fermentation and proof. (Reproduced with permission from Bloksma, A. H. 1990. *Cereal Foods World* 35:237–244.)

processing. Small strain dynamic rheological tests have certain advantages, such as their nondestructive nature, the simultaneous measurement of elastic and viscous moduli, and a well-developed theoretical background.

However, the deformation conditions do not correspond to those experienced by a dough during processing. Dynamic measurements mainly employ shear, whereas processing of dough (sheeting, proofing, and baking) is essentially biaxial extension. Rates of expansion during proof and oven baking are calculated to be in the order of 10^{-2}–10^{-4}/sec compared to those normally used in dynamic oscillatory tests, which may be several orders of magnitude greater. Also, dynamic measurements usually operate in the linear region at small strains of about 1%, while the strain in gas cell expansion during proof and oven rise can be in the order of several hundred percent—strains that are not attainable in shear.

Dynamic oscillation measurements are usually made over a relatively narrow range of frequencies, which are not relevant to the strain rates that occur during dough processing. Consequently, these measurements are not able to reflect the entanglement interactions of large glutenin molecules that, as we will see, are crucial to understanding dough behavior. It is therefore not surprising that no convincing relationship between dynamic rheological measurements and baking performance has ever been reported. The general picture that has emerged is that large strain extensional rheology is the most appropriate approach for gaining understanding of processes occurring during dough expansion (Safari-Ardi and Phan-Thien 1998; Dobraszczyk et al. 2003).

Strain hardening as a criterion for dough requirements

In recent times, strain hardening of dough as a criterion for good bread-making performance has emerged (van Vliet et al. 1992; Dobraszczyk et al. 2003). Dough expansion around gas cells causes thinning of gas cell walls. If the gas cell wall continues to expand along the thin region (see Figure 7.3), it may rupture. However, if the stress increases more than proportionally to the strain in the thin region, the thin region is stabilized against further deformation and the gas cell continues to expand along the thicker part of its wall.

This localized increase of stress in response to strain—resisting failure of gas cell walls—is called strain hardening. The tendency for the gluten–starch matrix to strain-harden depends on the dough properties,

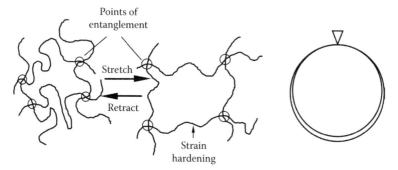

Figure 7.3 Proposed molecular mechanism of strain hardening (left) and the thinning of a region of the protein–starch matrix around the gas cell where strain hardening may act to stabilize the cell (right).

the force applied, and its rate of application. It has been shown that gas cell wall failure can be predicted by the Considere criterion for instability to extension for polymers (Dobraszczyk and Roberts 1994; McKinley and Hassager 1999; Wikstrom and Bohlin 1999). This criterion states that there is a critical strain value for every dough at a given strain rate and that, if this is exceeded, the dough wall will continuously thin and rupture. The force at any point of thinning is given by

$$F = \sigma A \tag{7.2}$$

where
F = applied force
σ = true stress at the point of thinning
A = area of cross section at the point

From Equation 7.2 it can be seen that, as the cross-sectional area A in the thin region is decreased, the stress will increase, causing continuous thinning of the region. However, as the force in this region increases more than the force in the thick region, the thin region is protected from continuous thinning or failure. If the force in the thin region decreases, rupture of the gluten–starch matrix becomes inevitable (Dobraszczyk and Roberts 1994). Strain hardening is represented by a parabolic relation between stress and Hencky strain, as shown in Figure 7.4 (Dobraszczyk 1999). The stress–strain curve of the gluten–starch matrix follows a power law equation:

$$\sigma = K\varepsilon^n \tag{7.3}$$

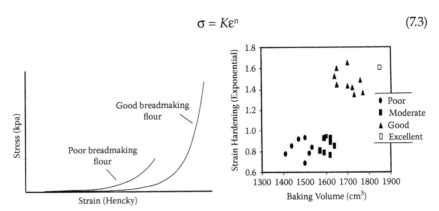

Figure 7.4 Stress–strain relationships for dough of good and poor bread-making flours (left) and the relationship between strain hardening index and baked loaf volume (right). (Data from Dobraszczyk, B. J. 1999. In *Bubbles in food*, ed. G. M. Campbell, C. Webb, and S. S. Pandiella, 173–182. St. Paul, MN: AACC International.)

where
σ = stress
K = constant
ε = Hencky strain
n = strain hardening index

The higher the value of *n*, the greater the strain hardening will be and the more likely it is that the bubble walls will be stable. This is reflected in the higher baked volumes for the flour with higher strain hardening index (Figure 7.4).

The relevance of strain hardening to dough expansion requirements has been advanced by the development of an apparatus (Dobraszczyk and Roberts 1994) employing a bubble inflation attachment for use with a texture analyzer (TA-XT2Plus). This has enabled stress–strain behavior of dough to be measured using large deformation biaxial extension—the type of extension that occurs in a leavened dough system. Some results using this system are shown in Figure 7.5.

Stress is plotted as a function of Hencky strain for flour from two wheat varieties of good (Hereward) and poor (Riband) bread-making performance at different temperatures. Both strain hardening index and failure strain (strain at which dough cell walls rupture) are greater for dough from Hereward flour than from Riband when compared at the same temperature. It is also evident that both these parameters decrease as temperature is increased and that Hereward is less affected by temperature rise than

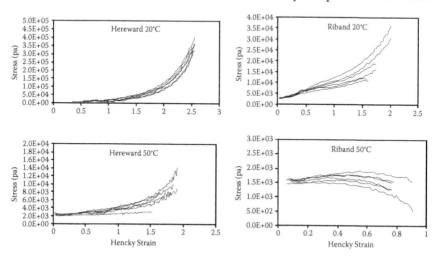

Figure 7.5 Stress–strain curves for two flours of good and poor bread-making performance at two temperatures. (Data from Dobraszczyk, B. J. et al. 2003. *Cereal Chemistry* 80:218–224.)

Riband. The strong relationship between failure strain and strain hardening index found by Dobraszczyk and Roberts (1994) is shown in Figure 7.6.

This relationship appears to be curvilinear and suggests that strain hardening index is reaching a limiting value at the highest values of failure strain. The range of the relationship of failure strain to strain hardening index has been extended by examining artificial flours made by adding protein fractions to base flours (Sroan, Bean, and MacRitchie 2009). The protein fractions were prepared by a sequential pH fractionation of gluten (Gupta, Bekes, and MacRitchie 1990). Fractions obtained from the supernatant at the highest pH (5.3) were concentrated in monomeric proteins (gliadins) and, as pH was lowered, the ratio of polymeric/monomeric proteins increased in successive fractions. The composition of the fractions in terms of percentages of total polymeric protein (TPP) and unextractable polymeric protein (UPP) is summarized in Table 7.1.

Test-bake loaf volume and dough strain hardening index for two base flours with 1% addition of gluten protein fractions were plotted as a function of the percentage of the total polymeric protein that had been extracted up to each point. The graphs are shown in Figure 7.7. Two obvious conclusions

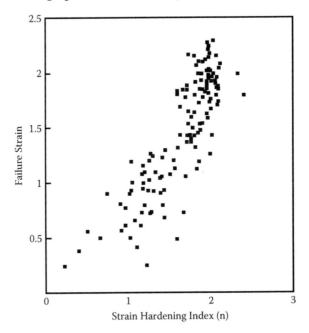

Figure 7.6 Relationship between failure strain and strain hardening index. (Reproduced with permission from Dobraszczyk, B. J., and C. A. Roberts. 1994. *Journal of Cereal Science* 20:265–274.)

Table 7.1 Total Polymeric Protein (TPP) and
Unextractable Polymeric Protein (UPP) in Gluten
Protein Fractions Obtained by pH Fractionation

Fraction	TPP (%)	UPP (%)
pH 5.3	31.6	23.9
pH 4.9	52.6	72.6
pH 4.1	64.1	78.3
pH 3.5	65.5	77.6
pH 3.1 supernatant	68.0	83.2
pH 3.1 residue	69.5	86.6

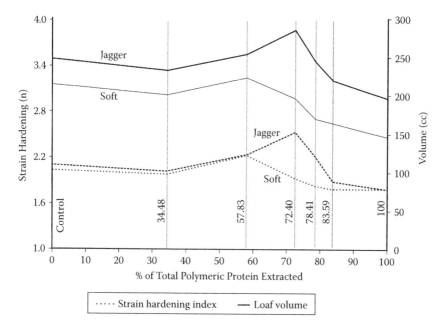

Figure 7.7 Strain hardening index and bake-test loaf volume as a function of the percentage of polymeric protein extracted from flour of variety Jagger. (Sroan, B. S. 2007. Ph.D. thesis, Kansas State University, Manhattan, KS.)

can be made from these results. First, each of the two dependent variables shows closely similar behavior. Second, the two variables decrease with addition of the pH 5.3 fraction (high in monomeric proteins), increase with successive fractions to a maximum, and then decline with addition of the later extracted fractions. Thus, each variable has an optimum protein composition. If the flour strength is raised above the optimum, both variables (loaf volume and strain hardening index) decrease.

Molecular interpretations of dough rheology

In keeping with the objectives of the book, we will try to explain the observed macroscopic properties of dough at a molecular level. The behavior shown in Figure 7.7 with respect to test-bake loaf volume and strain hardening index is instructive. What is being changed is the molecular weight distribution (MWD) of the gluten protein.

Strain hardening at a molecular level may be explained by the diagrams of Figures 7.3 (left) and 7.8. When a stress is applied to the glutenin

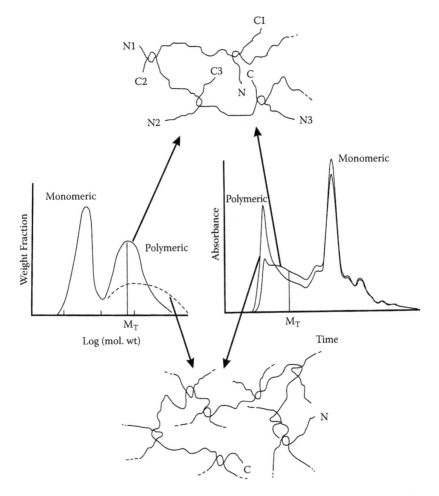

Figure 7.8 Schematic representation of two MWDs for polymeric protein. The lower one has an MWD shifted to higher values. It has a higher entanglement density and would be expected to have a higher strain hardening index. (From Sroan, B.S. 2007. Ph.D. thesis, Kansas State University. Manhattan, KS.)

network in the dough, chains between entanglements are stretched. After becoming fully stretched, further extension can only occur by slippage of chains through entanglements. Slippage through one entanglement puts stress on further entanglements along the molecule. The behavior then depends on the number of entanglements per molecule or the entanglement network density, which is determined by the MWD.

The extremes in behavior that can occur are illustrated by the stress–strain relationships shown in the schematic diagram of Figure 6.8 in the preceding chapter. The lower MWD imparted by addition of the pH 5.3 fraction decreases the entanglement network density, facilitating slippage of chains through entanglements and lowering the strain hardening index (n). This effect thus resembles the behavior of Figure 6.8b. As the MWD shifts to higher values with addition of fractions from progressively lower pHs, n increases until a maximum is reached. This corresponds to the behavior of Figure 6.8d for optimum conditions. Addition of later fractions increases the entanglement network density beyond the optimum, where slippage of chains through entanglements is not sufficiently rapid compared to the rate of extension to avoid breakage of chains (Figure 6.8c). This prevents a high strain hardening index from being attained. Failure strain and, as a result, loaf volume follow the same trend.

Secondary stabilizing mechanism

We saw in Chapter 5 that a separate liquid phase is present in dough. This liquid phase plays a crucial role in dough expansion. The initial concentration of gas nuclei and their size distribution created during dough mixing is determined by how they are incorporated in the liquid phase and by the surface active components that adsorb at the air–aqueous interface and stabilize them.

The main surface active compounds present are proteins and lipids. The requirement for surface activity is that the compound have molecules with a dual nature; that is, they have a polar moiety and a nonpolar moiety. Such molecules concentrate at an interface because they can orient so that the polar moiety interacts with the polar phase (water) and the nonpolar moiety interacts with the nonpolar phase (air). An example is shown in Figure 7.9. A molecule such as a long chain fatty acid has a lower energy at the interface than in either of the adjacent phases. It will therefore become concentrated and oriented at a polar–nonpolar interface.

The stabilizing liquid film surrounding gas bubbles in dough becomes important when bubbles come into close contact. As we have seen, this occurs when the relative dough volume attains a value around 3.85 (i.e., during the late proofing stage). The model of Figure 7.1 then assumes

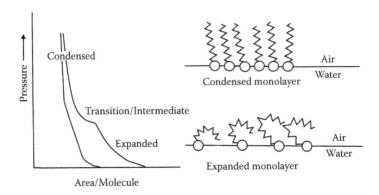

Figure 7.9 Surface pressure-area ∏-*A* relationships of monolayers showing the different phases. The molecular configurations for condensed and expanded monolayers are shown (right). Polar head groups are represented by circles and hydrocarbon chains by zig-zag lines.

particular importance. This model was proposed by Gan et al. (1990) based on electron microscopy of fermented dough. It was found that during late proofing, the starch–protein matrix surrounding gas cells could become discontinuous but that the bubbles continued to expand. This could be explained by the presence of thin liquid lamellae enveloping the gas cells, which prevented coalescence of the bubbles.

Many observations are consistent with this secondary stabilizing mechanism. It has been found, for example, that small amounts of flour lipids can have large effects on loaf volume in a baking test in which exogenous lipid additives were not included in the formulation (MacRitchie and Gras 1973). These compounds have no effect on dough rheology. The small amounts needed to have large effects on loaf volume can be rationalized by realizing that formation of monomolecular films at the gas–liquid interface requires only small amounts of material.

Some results for the effects of addition of different lipid compounds on loaf volume of a base flour are shown in Figure 7.10 (Sroan 2007). The natural lipid from flour has been separated into two fractions: polar and nonpolar. Thin layer chromatography (TLC) of the fractions is shown in Figure 7.11. Polar lipids comprise mainly galactolipids and phospholipids. Their addition enhances loaf volume. In contrast, the nonpolar fraction, which comprises mono-, di-, and triglycerides and free fatty acids as the main components, causes depression of the loaf volume. The unsaturated linoleic acid, which is the major fatty acid in wheat, also depresses loaf volume, whereas the saturated palmitic acid has no effect. Another saturated fatty acid, myristic acid, affected loaf volume negatively.

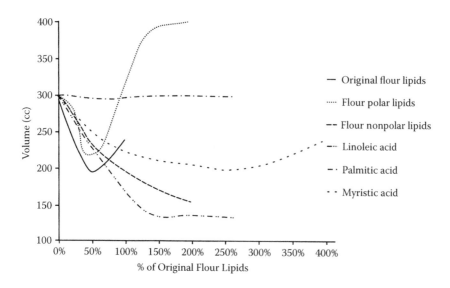

Figure 7.10 Loaf volume versus percentage of added flour lipid for different lipid fractions added to defatted flour from variety Jagger. (Sroan, B. S. 2007. Ph.D. thesis, Kansas State University, Manhattan, KS.)

Adsorbed monomolecular films

In order to understand the different effects of lipids on gas cell stability (Figure 7.10), we need to digress momentarily to focus on the properties of monomolecular films or monolayers. Monolayers may be studied by the use of a surface film balance, also called a Langmuir trough. When a compound is spread or adsorbs at an interface, it generally lowers the interfacial tension. The change in interfacial or surface tension (γ) in the case of an air–aqueous interface is called the surface pressure (Π) and is defined by

$$\Pi = \gamma_0 - \gamma \qquad (7.4)$$

where
   γ_0 = surface tension between air and the pure aqueous phase
   γ = surface tension in the presence of a monolayer

Surface pressure has units of millinewtons per meter; that is, it is the two-dimensional analogue of three-dimensional pressure that has units of millinewtons per square meter. When different compounds are studied by the film balance, they display a range of properties. In film balance studies, it is usual to spread a known amount of a relatively insoluble surface-active compound from solution onto a clean surface. The surface area can then be

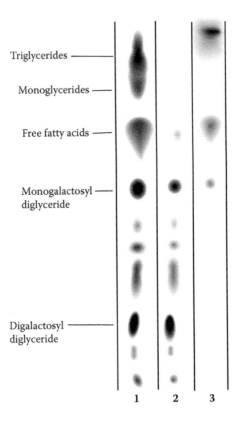

Figure 7.11 TLC patterns for (1) total flour lipid and its (2) polar and (3) nonpolar fractions. Polar lipids comprise phospholipids and galactolipids. The nonpolar fractions include mono-, di-, and triglycerides and free fatty acids.

varied by compressing a barrier, and an isotherm of Π versus A, the area per molecule, is measured. The range of Π-A relationships is shown in Figure 7.9. Analagous to three-dimensional systems, which exist in three main states of matter (vapor, liquid, solid), monolayers have three main two-dimensional states: gaseous (areas/molecule higher than expanded) expanded, and condensed, as shown in Figure 7.9.

Effects of lipids on loaf volume

It is generally found that good stabilization of foams and emulsions requires compounds that give condensed monolayers. Compounds that give expanded or gaseous monolayers do not, as a rule, act as stabilizers; in fact, they are more likely to be destabilizers. The compounds that enhance loaf volume or have little effect are those that give condensed monolayers at room temperature (i.e., the galactolipids and phospholipids of the polar lipid fraction and the saturated fatty acid, palmitic acid).

The phase change from condensed to expanded monolayer can occur in a number of ways (e.g., by increasing the temperature, increasing the degree of unsaturation in a molecule, or decreasing its hydrocarbon chain length). The unsaturated linoleic acid (18:2) and the short chain myristic acid (14:0) both depress loaf volume. De Stefanis and Ponte (1976) demonstrated that linoleic acid was the component of nonpolar lipid that contributes to deleterious effects on loaf volume and that palmitic acid had no negative effect.

Barrier to coalescence

The literature on theories that have been proposed to explain the mechanism by which surfactants stabilize dispersed systems such as foams and emulsions is vast. Here, we will briefly mention one approach that can be related to the loaf volume behavior depicted in Figure 7.10 (MacRitchie 1976, 1990).

In most systems, stable bubbles can only be formed in water or an aqueous solution if a surface active agent is present at the interface. When two bubbles are brought into contact, the resistance to coalescence will be determined by the nature of the surfactant film or monolayer. The result of coalescence is always a decrease of interfacial area. However, before coalescence can occur, the surface films around the bubbles are compressed. This gives an increase in Π and provides an elastic restoring force tending to oppose the compression.

Suppose we assume that a certain small area of hole needs to be created between the two bubbles for coalescence to proceed spontaneously; that is, the areas of the two monolayers must be reduced by a certain amount. The magnitude of the restoring force depends on the compressibility of the monolayer. In the case of a condensed film, a high value of Π will be obtained, giving rise to a correspondingly large restoring force resisting further compression. An expanded monolayer gives a much lower increase of Π for the same decrease of area and thus a lower restoring force. The compound giving the expanded monolayer may also be slightly soluble and will tend to desorb, thus "short-circuiting" the surface pressure barrier.

Demonstration

Three test tubes containing about 10 cm of pure water are provided. The first one is vigorously shaken. No foam is produced (if some bubbles are seen, this shows that the water is not pure and contains a minute amount of surface active compounds). A small amount of a solid surfactant (e.g., sodium dodecyl sulfate) is added to the second test tube and shaken. A stable foam is produced. This is repeated with the third test tube. Some alcohol (e.g., propanol) is gently poured down the side of the third test tube. The foam will disappear. This simple experiment is intended to draw attention to the important role of surfactants in stabilizing gas–liquid interfaces in foams. The sodium dodecyl sulfate and the propanol illustrate the actions of a foam stabilizer and destabilizer, respectively.

Exercises

1. In the study of the effects of wheat nonpolar lipids on bread-making, deleterious effects were found to be caused by free fatty acids. Within the fatty acids, detrimental effects were related to linoleic acid but not to palmitic acid. Could the difference be due to different behavior of these fatty acids at the gas–liquid interface?

2. Long chain saturated alcohols with even numbers of carbon atoms give condensed monolayers whose Π-A curves extrapolate to 20 ± 0.25 Å^2 at 20°C. It is found that 36 μL of a 1.0-mg/mL solution of an unknown alcohol spreads to give a monolayer whose Π-A curve extrapolated to an area of 162.8 cm^2. Deduce the identity of the alcohol.

References

Bloksma, A. H. 1990. Dough structure, dough rheology and baking quality. *Cereal Foods World* 35:237–244.

De Stefanis, V. A., and J. G. Ponte. 1976. Studies on the bread-making properties of wheat-flour nonpolar lipids. *Cereal Chemistry* 53:636–642.

Dobraszczyk, B. J. 1997. Development of a new dough inflation system to evaluate doughs. *Cereal Foods World* 42:516–519.

———. 1999. Measurement of biaxial extensional rheological properties using bubble inflation and stability of bubble expansion in bread doughs. In *Bubbles in food*, ed. G. M. Campbell, C. Webb, and S. S. Pandiella, 173–182. St. Paul, MN: AACC International.

Dobraszczyk, B. J., and C. A. Roberts. 1994. Strain hardening and dough gas cell-wall failure in biaxial extension. *Journal of Cereal Science* 20:265–274.

Dobraszczyk, B. J., J. Smewing, M. Albertini, G. Maesmans, and J. D. Schofield. 2003. Extensional rheology and stability of gas walls in bread doughs at elevated temperatures in relation to bread-making performance. *Cereal Chemistry* 80:218–224.

Gan, Z., R. E. Angold, M. R. Williams, P. R. Ellis, J. G. Vaughan, and T. Galliard. 1990. The microstructure and gas retention of bread dough. *Journal of Cereal Science* 12:15–24.

Gupta, R. B., F. Bekes, and F. MacRitchie. 1990. Functionality of glutenin, gliadin and secalin fractions as measured by extensograph and mixograph. In *Gluten proteins 1993*, 550–559. Detmold, Germany: Association of Cereal Research.

MacRitchie, F. 1976. Monolayer compression barrier in emulsion and foam stability. *Journal of Colloid Science* 56:53–56.

———. 1990. *Chemistry at interfaces*, 244–248. San Diego, CA: Academic Press.

MacRitchie, F., and P. W. Gras. 1973. The role of flour lipids in baking. *Cereal Chemistry* 50:292–302.

McKinley, G. H., and O. Hassager. 1999. The Considere condition and rapid stretching of linear and branched polymer melts. *Journal of Rheology* 43:1195–1212.

Safari-Ardi, M., and N. Phan-Thien. 1998. Stress relaxation and oscillatory tests to distinguish between doughs prepared from wheat flours of different varietal origin. *Cereal Chemistry* 75:80–84.

Sroan, B. S., S. R. Bean, and F. MacRitchie. 2009. Mechanism of gas cell stabilization in breadmaking. I. The primary gluten-starch matrix. *Journal of Cereal Science* 49:32–40.

Sroan, B. S. 2007. Mechanism of gas cell stability in breadmaking. Ph.D. thesis, Kansas State University, Manhattan, KS.

Van Vliet, T., A. M. Jansen, A. H. Bloksma, and P. Walstra. 1992. Strain hardening of dough as a requirement for gas retention. *Journal of Texture Studies* 23:439–460.

Wikstrom, K., and L. Bohlin. 1999. Extensional flow studies of wheat flour dough. II. Experimental method for measurements in constant extension rate squeezing flow and application to flours varying in bread-making performance. *Journal of Cereal Science* 29:227–234.

chapter eight

Shelf life

Introduction

The sequence of topics until now has been the milling of grain to flour, formation and properties of dough, and aspects of its processing. The next step in this logical order is to consider how the product maintains its quality over time (i.e., its shelf life). Again, we will direct the discussion to consider the deterioration of aerated products, specifically Western style bread. A number of staling processes may set in after the bread is baked (e.g., loss of crust crispness, changes in taste and aroma, and microbial spoilage). The focus in this chapter will be on the firming of the crumb, an inherent process that begins immediately on removal from the oven. Baked bread, fresh from the oven, is soft and attractive to consumers but loses its appeal relatively soon. Economic losses result and researchers have addressed the problem of crumb firming to try to understand the process and seek ways to eliminate it or at least to retard it.

Baked bread contains about 38% moisture and rolls contain somewhat less. Loss of moisture, of course, will cause firming of crumb, but early experiments showed that the process proceeds when moisture is preserved (Boussingault 1852). A number of theories of crumb firming have been proposed (see reviews by Axford, Colwell, and Elton 1966; Cauvain and Young 1998; Chinachoti and Vodovotz 2000; Gray and BeMiller 2003). The one that is presently favored and will be considered in this chapter is based on starch recrystallization (or retrogradation). However, in keeping with the ideals of scientific enquiry, this theory should be critically examined and readers should always remain open to alternative explanations. Previous chapters have highlighted protein in relation to dough properties and lipids with respect to processing aerated products. Here, we will be concerned primarily with the starch component. It is fitting to give prominence to starch because this is by far the most abundant component of cereals.

Starch composition and structure

Starch exists as granules in cereals. In wheat, it is usual to consider three classes of granules based on size: A (10–40 μm in diameter, lenticular in shape), B (1–10 μm diameter, roughly spherical), and C (<1 μm). Starch is a polymer of D-glucose and has two main components. Amylose is a linear molecule in which glucose units are linked through α, 1–4 bonds. Amylopectin is a branched molecule having α, 1–4 bonds, but with branching through α, 1–6 bonds. Amylose and amylopectin structures are shown in Figure 8.1. Molecular weights of amylose are in the hundreds of thousands and amylopectin molecular weights are thought to range in the millions. The ratio of amylopectin to amylose is about 70:30 in a normal wheat; however, in some genetic variants, the amylose content may be much greater or much less, down to close to zero (waxy wheats).

Starch gelatinization and retrogradation

When an aqueous starch suspension is heated, the granules absorb water and swell. If the temperature is below about 50°C, the process is reversible; however, above that temperature, a melting phenomenon referred to as gelatinization occurs. Because starch is not a pure compound in the chemical sense (linear and branched molecules, a range of molecular weights), melting occurs over a temperature range. When the suspension is cooled, a recrystallization, called retrogradation, proceeds.

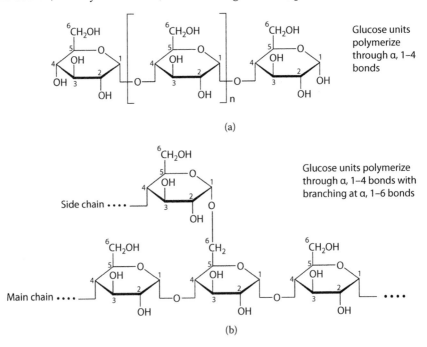

Figure 8.1 The structure of (a) amylose and (b) amylopectin.

Again, because starch does not consist of homogeneous molecules, the recrystallization differs from that of a pure compound. It resembles the picture shown in Figure 8.2. Segments of polymer chains align to form crystallites so that crystalline and amorphous segments are interspersed. In a system such as bread, the mobility of starch chains is low. The process of retrogradation can therefore be quite slow due to the need for polymer segments to diffuse and align. Starches from different cereals vary in their granular size and also in the temperature range of their gelatinization.

Characteristics of bread firming

In texture analysis, softness and firmness express the same concept inversely, although they are not simple reciprocals. Softness is defined as the deformation occurring at constant load and is normally what is measured qualitatively by a consumer to gauge freshness of bread. Firmness is the force required to provide a constant deformation; it is the measurement usually made to assess the progress of crumb firming of bread using an instrument such as the Baker compressimeter or the

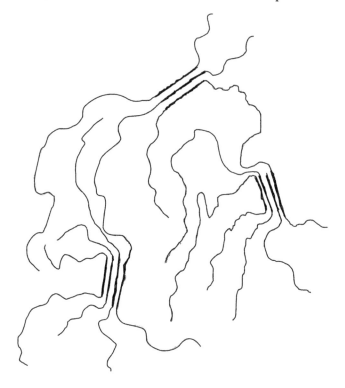

Figure 8.2 Crystallization of a heterogeneous polymer (e.g., starch).

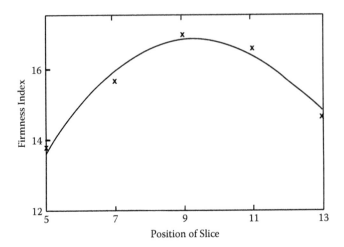

Figure 8.3 Pattern of firmness within a loaf of bread. (Reproduced with permission from Short, A. L., and E. A. Roberts. 1971. *Journal of the Science of Food and Agriculture* 22:471–472.)

more modern texture analyzer TA-XT2. Firmness of bread is taken as the force required to compress a bread slice by a certain percentage of its thickness (often 25%). When firmness measurements are made on a loaf of bread, it is important to be aware of how the pattern of firmness changes within a loaf.

An example of this pattern is shown in Figure 8.3 (Short and Roberts 1971). Firmness was measured on slices of bread after 3 days' storage at 5°C. This bread was made from 400 g of flour, but the same pattern was obtained for other weights of flour (300 and 500 g). Firmness was highest in the central slice (slice 9) and decreased as the slices approached the ends of the loaf. The firmness index (defined as the force to compress a slice by 30% of its thickness) could be approximately fitted to a parabola of the following form:

$$Y = A - B (x - c)^2 \tag{8.1}$$

where
  Y = expected firmness
  x = position of the slice
  c = position of the slice with maximum firmness
  A = maximum firmness
  B = rate of change of firmness with change in position

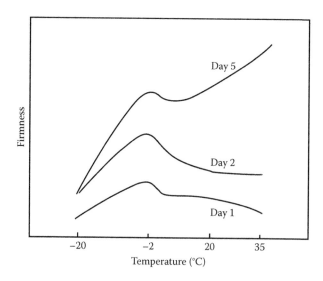

Figure 8.4 Firmness of bread for different times after baking as a function of storage temperature. (Reproduced with permission from Marston, P. E., and A. L. Short. 1969. *Food Technology in Australia* 21:154–159.)

Effect of temperature on firming rate

The regression coefficients for firming of bread at different temperatures (approximately equal to *B* in Equation 8.1) have been measured after several storage times and the results are plotted in Figure 8.4 (Marston and Short 1969). The maximum rate of firming is found to occur at a temperature close to (but not at) the freezing point of water. As the storage temperature moves away from this temperature on both sides, the rate of firming (regression coefficient) decreases. At very low temperature (freezer temperature, −20°C), firming is practically undetectable. At sufficiently high temperature (above 50°C), it also becomes very low, although it is more difficult to measure experimentally.

What could be an explanation for the unusual behavior on day 5 at the higher temperatures? After bread has firmed during storage, it can be freshened by reheating. However, with successive cycles of cooling and reheating, it is found that retention of freshness is reduced after reheating. It should be noted that these experiments are difficult to carry out while maintaining conditions (e.g., moisture content) at the same levels.

Theory of crumb firming

The observations that have been made, some of which have just been described, are mostly consistent with an explanation of crumb firming based on the partial melting of starch during baking and its subsequent slow recrystallization after cooling. The refreshening of bread by heating (albeit partial) is explained by the melting of retrograded (partially recrystallized) starch molecules. It is logically presumed that crystalline starch contributes to rigidity and thus to crumb firming. The amylopectin component is supposed to be involved in the process because the melting temperature of amylose is believed to be higher than the temperatures reached during baking.

The kinetics of crumb firming have been described by an equation derived from the Avrami theory (Avrami 1939). This theory describes the rate of change of a supercooled amorphous material to an ordered crystalline structure when the process is governed by random production of stable nuclei:

$$\theta = (E_1 - E_t)/(E_1 - E_0) = \exp(-kt) \tag{8.2}$$

where
θ = fraction of uncrystallized material
E_1 = limiting value of the crumb modulus
E_t = modulus after time t
E_0 = initial modulus
k = rate constant

Crystallization of polymers

The theory of crystallization of polymers is currently accepted as providing a model for crumb firming in bread. Two separate processes are involved in crystallization: nucleation and crystal growth. Initially, it is necessary for nuclei to form. A nucleus (or embryo) from which a crystallite can grow must consist of a small region in which the local arrangement of polymer chain segments happens to resemble the fixed structure of the crystallite, simply as a result of random movements of chains. These nuclei are continually being formed, but there is a critical size below which they decompose and above which they become stabilized by crystal growth.

The reason for the instability below a certain size is an increase in surface free energy that outweighs the decrease resulting from the phase change. (The concept of free energy is discussed in Chapter 9.) At temperatures just below the melting point, the rate of formation of nuclei is small because of the small difference in free energy between the crystalline and amorphous states. As the temperature is lowered, this difference becomes greater and the rate of nucleation increases correspondingly.

The second process is crystal growth, which essentially depends on the diffusion of segments from the melt to the crystal-melt interface. At temperatures well below the melting point, the growth rate of the nuclei, rather than their rate of formation, determines the kinetics of crystallization. This growth rate is normal in that it diminishes with falling temperature. As a result, the rate of crystallization does not increase indefinitely as the temperature is lowered but rather passes through a maximum.

As the temperature is lowered further, a point is reached where many physical properties of the polymer, such as elastic moduli, viscosity, and specific volume, fall dramatically over a narrow temperature range. This is the glass transition temperature, T_g, which was discussed in Chapter 6. At this temperature, segmental motions of polymer molecules practically cease. It appears that the temperature of around −20°C, at which crumb firming rate becomes negligible, may correspond closely to the T_g of amylopectin at the water content of bread.

Nucleation

In a supersaturated system, embryo nuclei are continuously being formed, but their formation is necessarily accompanied by the creation of surface free energy. Up to a certain nucleus size, the contribution of this positive free energy term is increasing faster than the negative contribution resulting from the formation of a new phase. Once the nucleus surpasses this critical size, however, the trend is reversed and the negative contribution progressively outweighs the positive one of surface energy. At this point, the nucleation process becomes spontaneous.

Nucleation is an example of the more general theory of reaction rates (Glasstone, Laidler, and Eyring 1941) depicted in Figure 8.5. In order for a reaction to occur, the reactant must attain a free energy state corresponding to the maximum in Figure 8.5 before it can transition to the product. This state is referred to as the activated or transition complex. The rate of a reaction is then proportional to the rate of formation of activated complexes. For the nucleation process, if r is the radius of the embryo nucleus in a supersaturated system, then the total change in free energy (ΔG) is given by the sum of the two contributions of opposite sign:

$$\Delta G = [-(4/3\,\pi\,r^3)/v]\cdot(G_c - G_a) + 4\,\pi\,r^2\gamma \qquad (8.3)$$

where

v = molar volume of the polymer

$G_c - G_a$ = molar free energy difference between the polymer in the crystal and the amorphous melt

γ = surface tension of the crystal embryo

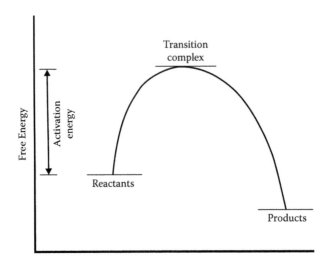

Figure 8.5 Graphical representation of activation energy concept. Reactants must attain a free energy state corresponding to the transition complex before reaction can proceed.

The function ΔG initially increases, passes through a maximum at the critical radius, and thereafter decreases. When values of the different terms in Equation 8.3 are known and ΔG is differentiated with respect to r, the critical radius of an embryo can be obtained from the condition that the derivative is zero at the maximum (i.e., $d\,\Delta G/dr = 0$). The change in the free energy of crystallizing starch is illustrated graphically in Figure 8.6. The activated complex in nucleation is a nucleus with a radius equal to the critical radius (r_{crit}).

The crystallization behavior of starch and thus the firming of bread are typical of polymer crystallization. For example, the rate of crystallization of natural rubber is shown in Figure 8.7. The maximum rate of

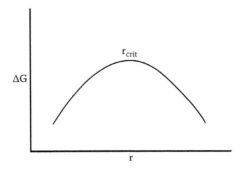

Figure 8.6 Change in free energy of a crystal nucleus as a function of its radius.

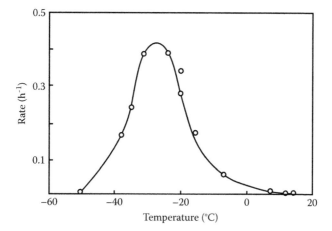

Figure 8.7 Rate of crystallization of natural rubber as a function of temperature. The ordinate is the reciprocal of the time required for one-half the total volume change. (Wood, L. A. 1946. In *Advances in colloid science*, vol. II, ed. H. Mark and G. S. Whitby, 57–95. New York: Interscience.)

crystallization occurs at a temperature intermediate between a lower temperature (possibly the glass transition temperature) and a higher temperature (the melting point) (Wood 1946).

Exercises

1. Make a list of observations about bread crumb firming that have been made.
2. We have seen in our discussion of dough properties in Chapter 5 that the properties of a dispersed system correspond to those of the continuous phase. How then do we explain the firming of bread crumb on the basis that starch is the dispersed phase?
3. Immediately after baking, bread contains a high concentration of water vapor in the gas phase. Tabulate the steps you would follow to calculate the thickness of the water film that would condense on the gas cell surface during cooling, assuming no net loss or diffusion of water.

References

Avrami, M. 1939. Kinetics of phase change. I. General theory. *Journal of Chemical Physics* 7:1103–1112.

Axford, D. W. E., K. Colwell, and G. A. H. Elton. 1966. *Bread staling: The current state of knowledge*, 1–19. Chorleywood, England: British Baking Industries Research Association.

Boussingault, J. G. 1852. Experiments to determine the transformation of fresh bread to stale bread. *Annales de Chimie Physique* 36:490–494.

Cauvain, S. P., and L. S. Young. 1998. *Technology of bread making,* 248–260. Gaithersburg, MD: Aspen Publishers.

Chinachoti, P., and Y. Vodovotz. 2000. *Bread staling.* Boca Raton, FL: CRC Press.

Glasstone, S., K. J. Laidler, and H. Eyring. 1941. *The theory of rate processes.* New York: McGraw–Hill.

Gray, J. A., and J. N. BeMiller. 2003. Bread staling: Molecular basis and control. *Comprehensive Reviews in Food Science and Food Safety* 2:1–21.

Marston, P. E., and A. L. Short. 1969. Factors involved in the storage of bread. *Food Technology in Australia* 21:154–159.

Short, A. L., and E. A. Roberts. 1971. Pattern of firmness within a bread loaf. *Journal of the Science of Food and Agriculture* 22:471–472.

Wood, L. A. 1946. Crystallization phenomena in natural and synthetic rubbers. In *Advances in colloid science,* vol. II, ed. H. Mark and G. S. Whitby, 57–95. New York: Interscience.

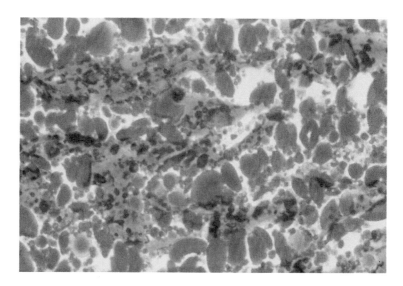

Figure 5.7 Photomicrograph of dough in the early stage of mixing. Starch has been stained by Ponceau 2R; starch granules are in red dye and protein phase in blue-green. (Courtesy of R. Moss, Bread Research Institute of Australia, North Ryde, NSW, Australia.)

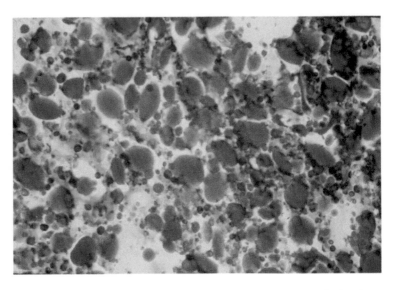

Figure 5.8 Photomicrograph of dough at the later stage of dough development. Starch has been stained by Ponceau 2R; starch granules are in red dye and protein phase in blue-green. (Courtesy of R. Moss, Bread Research Institute of Australia, North Ryde, NSW, Australia.)

chapter nine

Solubility of cereal proteins

Introduction

Cereals contain a mixture of different protein classes. The earliest classification by Osborne (1907) separated the proteins into four main groups based on solubility: albumins (soluble in water), globulins (soluble in dilute salt solution), prolamins (soluble in aqueous ethanol), and glutelins (partially soluble in dilute acid and alkali and forming the residue after sequential extraction of the other groups). The Osborne classification has proved valuable for the development of cereal protein chemistry. It is difficult, however, to obtain a sharp separation of different protein groups based on solubility because different classes tend to overlap. As a result, recent techniques such as high-performance liquid chromatography (HPLC; discussed in Chapter 10) are able to give a sharper resolution of the fractions.

Some cereal proteins are quite soluble in aqueous solution (albumins, globulins) but the functional proteins—prolamins and glutelins—are difficult to solubilize. This is important because, for example, we do not want wheat gluten proteins to be soluble in a dough system. However, in order to characterize proteins, it is usually necessary to have them in solution. Because the property of solubility is such an important one with respect to cereal proteins, we will dedicate this chapter to the topic. In keeping with the objectives of the book, we will attempt to approach the subject from the most general and fundamental basis.

Why do some compounds dissolve in a solvent but others do not? This is part of a more general question: Why do some physical processes (such as solubilization) or chemical reactions proceed while others do not, and why do they stop at a certain stage? Let us first try to answer this question and then consider how it applies to an understanding of solubility.

Potential energy and mechanical processes

In purely mechanical systems, a basic law governs behavior. The state of equilibrium is the position of minimum potential energy, and everything that happens is an approach to this state. For example, a book falling from a height to a table represents an approach to a minimum of potential energy.

All physical processes and chemical reactions are an approach to some equilibrium state. The question arises: Does this same principle of minimum potential energy apply to chemical reactions or physical processes?

First, we need to know how potential energy can be measured in a reaction. Energy is always conserved (a result implicit in the first law of thermodynamics), so if it is lowered, the extra energy must be given off as heat, which will cause the surroundings to warm up. This amount of heat can be measured accurately by a calorimeter and is defined as the enthalpy (H). It is made up of two terms: the change in internal energy, E, of the molecules and any energy or work of the form PV due to compression or expansion (change in volume, V) that has to be done against a pressure P. If this minimum potential energy condition applies, then all spontaneous reactions should be accompanied by an evolution of heat or, in other words, a decrease of enthalpy. Indeed, it is found that most reactions are exothermic (i.e., they give off heat).

However, some reactions occur spontaneously with an absorption of heat—that is, an increase of enthalpy. An example of such a reaction that absorbs a large amount of heat and therefore must go to a higher enthalpy state is given at the end of this chapter. In such endothermic reactions, we can presume that the potential energy increases. It can therefore be concluded that at least one other factor needs to be considered.

Chemical systems and the tendency for disorder

A chemical system is different from a purely mechanical one in that it contains a very large number of discrete particles—the constituent atoms or molecules of the substances. When we study systems with large numbers of particles, another universal tendency becomes apparent: the tendency to become mixed up or disordered (a principle embodied in the second law of thermodynamics). For example, if two gases are brought into contact, they mix together by diffusion and/or convection to form a homogeneous mixture. In so doing, they go from an ordered state to one of maximum disorder. A simple practical example to illustrate this is given at the end of the chapter. A measure of the degree of disorder in such a mixing process has been called the entropy (S).

We thus see that two basic drives cause a process/reaction to proceed toward equilibrium:

1. The enthalpy tends to a minimum.
2. The entropy tends to a maximum.

The question then arises as to what happens in a real situation when enthalpy and entropy are forced to compromise.

The concept of free energy

In order to minimize the effects of variables, scientists usually prefer to conduct experiments at constant pressure and temperature. When this is done and only the system being studied is considered, a new function—called the Gibbs free energy (hereafter called the free energy)—has been introduced. It is defined as follows:

$$G = H - TS \tag{9.1}$$

where
G = free energy (joules)
H = enthalpy or heat content (joules), which is essentially a measure of the potential energy
S = entropy (joules/degree Celsius)
T = absolute temperature (kelvins)

The free energy is a concept that embodies the two universal drives of minimization of enthalpy and maximization of entropy. As the term suggests, it is the energy that is available for performing useful work, such as making a process proceed. A system can proceed spontaneously from one state to another only if accompanied by a lowering of free energy. An alternative way of expressing this is that the condition of equilibrium, under conditions of constant temperature and pressure, is that the free energy of the system be a minimum.

Free energy relations for the water system

To illustrate the principle of free energy, let us consider how it applies to a water system. In the idealized diagram of Figure 9.1, the free energy of 1 mol (18 g) of water is plotted as a function of temperature for each of the three states—solid, liquid, and vapor. Equation 9.1 is a linear equation as long as H and S remain constant. This condition is approximately true over a moderate temperature range, as depicted in the schematic diagram of Figure 9.1. The slopes of the lines give the molar entropies and the intercepts at $T = 0$ give the molar enthalpies.

In going from ice to liquid water and then to water vapor, the slope of the line becomes greater because we are changing from ordered to more disordered states (i.e., higher entropies). The enthalpy of liquid water is greater than that of ice by an amount equal to the heat of fusion (AC). This reflects the increase in potential energy of molecules when the bonds in ice are broken. A greater increase in potential energy is attained when the liquid vaporizes. Thus, CE (the distance between points C and E on the y-axis of Figure 9.1) is equal to the heat of vaporization of water.

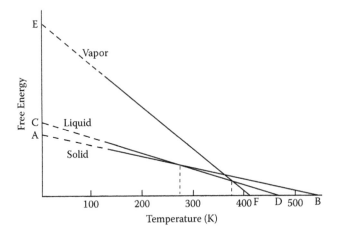

Figure 9.1 Schematic diagram of the free energy of 1 mol of water as a function of temperature at 1 atm pressure. (MacRitchie, F. 1990. *Chemistry at interfaces.* San Diego, CA: Academic Press.)

The most stable state (the state of equilibrium) at any temperature is the state of lowest free energy. It can be seen that, at two temperatures (273 and 373 K at standard pressure), two phases have the same free energy and are therefore in equilibrium. These are the freezing point and boiling point of water.

Entropy and probability

It is relatively easy in theory to measure changes in enthalpy by monitoring the heat changes accompanying a process or reaction. Let us examine more closely the concept of entropy to see how that can be characterized quantitatively. A perfectly ordered system in which two separate layers of unmixed gases are in contact is a highly improbable state. As a result of thermal energy, it will evolve toward a system in which the two gases are homogeneously mixed—the state of maximum disorder. On reaching this state, the probability of reverting spontaneously to the initial ordered state is not impossible; however, the probability of it happening is infinitesimally low.

The thermodynamic probability of a system is defined as the ratio of the probability of an actual state to one of the same total energy and volume in which the molecules are completely ordered. This suggests that entropy is a function of probability (P); that is,

$$S = f(P) \tag{9.2}$$

However, the entropy of two systems is equal to the sum of the entropies of the individual ones; that is, it is additive: $S = S_1 + S_2$. On the other hand, the probabilities of two independent individual events (P_1 and P_2) are multiplied together to obtain the probability of the combined event; that is,

$$P = P_1 P_2 \tag{9.3}$$

The only relation that satisfies both Equations 9.2 and 9.3 is a logarithmic one:

$$S = k \ln P \tag{9.4}$$

This relationship has been well confirmed and the value of the constant k (Boltzmann's constant) has been found to be 1.38×10^{-23} J/degree.

Let us see how Equation 9.4 can be applied to give a measure of entropy. When we work with systems containing large numbers of particles, the probability is taken as the number of ways of arranging the particles (W) so that $S = k \ln W$. Suppose we take a simple example of three molecules of A and three molecules of B and we calculate the number of ways of arranging them at a surface in a close-packed arrangement. If the molecules of each substance are of similar size, we can use a two-dimensional lattice to work out the number of possible arrangements. Probability theory gives the number as

$$W = (3 + 3)!/3! \cdot 3! = 20 \tag{9.5}$$

The numerator gives the total number of possible arrangements and the denominator corrects for those arrangements that are indistinguishable because molecules of A are identical, as are molecules of B. The 20 arrangements are illustrated in Figure 9.2. The change in entropy as a result of mixing the three molecules of each substance is given by $S = k \ln 20$. This is a very small number. However, when we work with molar quantities, we are dealing with numbers of particles in the order of 10^{23}.

The value of W then becomes large and the increase in entropy of mixing (and therefore the decrease in free energy) is appreciable. By applying Equation 9.5 to molar quantities and assuming that the molecules of solute and solvent (a and b, respectively) are the same size, we can arrive at a relatively simple general expression (not derived here) for the entropy of mixing (S^M):

$$S^M = -R\, x_a \ln x_a - R\, x_b \ln x_b \tag{9.6}$$

where x_a and x_b are the mole fractions of a and b.

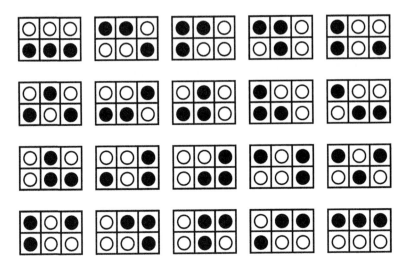

Figure 9.2 The 20 possible arrangements for three molecules of *A* and three molecules of *B* in a two-dimensional lattice. (MacRitchie, F. 1990. *Chemistry at interfaces.* San Diego, CA: Academic Press.)

The entropy increase as a result of mixing is a major driving force for dissolution. For some solutions, there is no change of enthalpy on mixing solute and solvent (athermal solutions) and, if molecules of solute and solvent are of similar size and shape, the lowering of free energy is given exactly by the entropy of mixing multiplied by the temperature (K). Such solutions are called perfect solutions. The solute and solvent should be miscible in all proportions. When substances are immiscible (i.e., when it is difficult to dissolve a solute in a solvent), this arises because of deviations from perfect behavior. The deviations may have their origin in enthalpic effects, entropic effects, or a combination of both.

Enthalpic and entropic effects on molecular properties

The preceding theory provides a fundamental foundation for understanding the solubility properties of proteins. However, to gain a greater understanding, we need to invoke molecular theory. Let us look at two simple examples where we can combine thermodynamic principles with molecular properties.

Figure 9.3 shows a dissolved polymer molecule that is subjected to a tensile force. Let us suppose that enthalpy does not change when the molecule is introduced into the solvent. Its configuration will therefore be that of maximum entropy, a perfectly random coil. On being stretched,

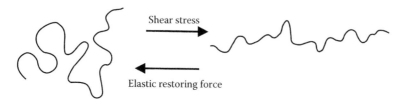

Figure 9.3 Schematic illustration of the elastic restoring force for a polymer molecule after stretching. (MacRitchie, F. 1990. *Chemistry at interfaces.* San Diego, CA: Academic Press.)

the molecule becomes more ordered and its entropy decreases. This imparts an elastic restoring force, causing the molecule to try to return to its original configuration of lowest free energy. This is the type of force that largely explains rubber elasticity. Of course, this will be modified by any enthalpic effects that may be present.

The other example, shown in Figure 9.4, illustrates how the conformation of a polymer molecule is affected by its interaction with solvent molecules. When heat does not change on introduction of the polymer into the solvent ($\Delta H = 0$), the polymer molecule adopts the configuration of highest entropy, a random coil. If ΔH is positive (i.e., enthalpy of the system increases when polymer and solvent are mixed), the polymer molecule will try to minimize interactions with solvent molecules by adopting a folded configuration.

However, folding incurs an increase of order, which will be resisted. The final configuration will therefore be a compromise between the two driving tendencies of decreasing enthalpy and increasing entropy. In the case of a negative enthalpy of solution (ΔH negative), the polymer will stretch out to try to maximize interactions with solvent, but, again, this leads to a more ordered conformation and a compromise has to be reached.

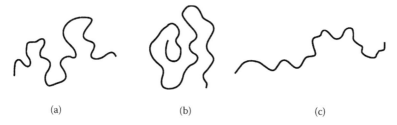

(a)   (b)   (c)

Figure 9.4 Schematic representations of the configuration of a polymer molecule with different values of ΔH for interaction between polymer segments and water molecules: (a) $\Delta H = 0$; (b) ΔH positive; (c) ΔH negative. (MacRitchie, F. 1990. *Chemistry at interfaces.* San Diego, CA: Academic Press.)

Solubility properties of proteins

Proteins consist of polypeptide chains, each with a specific sequence of amino acid side chains. The amino acid side chains (or residues if we include the peptide portion) can be placed in different classes according to their structure. The simplest classification on that basis is into polar and nonpolar classes. We will be dealing with solubility in aqueous solution, so the polar residues will tend to be soluble and nonpolar residues will tend to be insoluble.

Proteins have a balance between polar and nonpolar residues and this balance will largely determine the solubility. If all of the residues were nonpolar (e.g., polyleucine), we would expect the protein to be insoluble. Within the two main classes of residues, there are degrees of polarity (hydrophilicity) and nonpolarity (hydrophobicity). The most polar residues are the ionized ones, such as glutamic acid at high pH and lysine at low pH. Attempts have been made to use amino acid composition of proteins to predict solubility. As a result, several parameters have been introduced and we will briefly examine these.

Nonpolar side chain frequency

Nonpolar side chain frequency (NPS) is defined as the number of residues of tryptophan, isoleucine, tyrosine, phenylalanine, proline, leucine, and valine divided by the total number of residues in the molecule (Bigelow 1967). The list of nonpolar side chains may vary between workers; for example, alanine, glycine, cysteine, and methionine may be considered to be nonpolar.

Charged group frequency

Charged group frequency (CHF) is defined as the proportion of amino acid residues that are charged at about pH 6. It is thus equal to the number of aspartic acid, glutamic acid, histidine, lysine, and arginine residues divided by the total number of residues in the molecule (Bigelow 1967).

Polarity ratio

Protein molecules tend to fold in aqueous solution so as to bury their nonpolar residues in the interior of the molecule (internal shell) and concentrate their polar residues at the periphery (external shell), consistent with steric requirements (cf. Figure 9.4b). Polarity ratio (P) is defined as

$$P = V_e/V_i \tag{9.7}$$

where V_e = volume of the external shell and V_i = volume of the internal shell.

Values of P are calculated by using specific volumes for amino acid residues and assuming all polar residues to be in V_e and all nonpolar ones to be in V_i (Fisher 1964).

Average hydrophobicity

Possibly a more rigorous parameter than polarity is the average hydrophobicity ($H\Phi$), which is calculated from free energies of transfer of amino acid residues (i.e., the change in free energy when a residue transfers from a nonpolar phase to an aqueous phase). $H\Phi$ of a protein is then the sum of the hydrophobicities of its constituent amino acid residues divided by the total number of residues (Dunnill 1965; Iqbal and Verrall 1988).

Surface hydrophobicity

Properties of proteins such as solubility and chromatographic behavior depend on the residues that are accessible to the solvent. The nonpolar residues buried in the interior of the molecules should not directly affect these properties. Therefore, the surface hydrophobicity may be the more relevant parameter. Of course, the ideal model depicted by the polarity ratio is never attained. Because of the steric constraints and the inevitable adjacency of polar and nonpolar residues in the amino acid sequence, a proportion of nonpolar residues will be found at or close to the periphery of molecules. For proteins whose structures have been determined (e.g., by x-ray crystallography), it is possible to calculate the surface hydrophobicity. Some values are recorded in Table 9.1. Because of their heterogeneity, cereal proteins are not easily amenable to structure determination, so their surface hydrophobicities have not been calculated.

Table 9.1 Solubility-Related Parameters for Proteins

Protein	NPS	$H\Phi$ (KJ/ residue)	P	CHF	RSH ($\times 10^{-3}$ Å^2/M$_r$)
Glutenin	0.34	3.87	1.16	0.13	
Gliadin	0.39	4.19	1.07	0.08	
Myoglobin	0.32	4.56	1.12	0.34	23.0
Hemoglobin	0.35	4.51	0.87	0.27	33.8
Ovalbumin	0.34	4.64	0.92	0.24	41.8

Notes: NPS = nonpolar side chain frequency; $H\Phi$ = average hydrophobicity; P = polarity ratio; CHF = charged group frequency; RSH = relative surface hydrophobicity.

Solubility-related parameters for cereal proteins

Table 9.1 summarizes data for the solubility-related parameters discussed previously for several well-characterized proteins and compares them with corresponding values for the two main wheat gluten proteins: gliadin and glutenin. No trend is apparent in the parameters related to polarity or hydrophobicity. The hydrophobicity parameters for the wheat proteins are similar to those of the well-characterized proteins. It can therefore be tentatively concluded that the relative difficulty in solubilizing the wheat proteins is not because of greater proportions of nonpolar residues. However, there appears to be a dearth of charged amino acid residues and this may, at least in part, explain their low solubility. Table 9.2 gives similar data calculated for proteins from different cereals. Again, it appears that no trend is seen for hydrophobicity but charged group frequency is relatively low for cereal proteins.

Effect of molecular size

At least one other factor has an impact on solubility: the molecular size of the large glutelin proteins. Why are larger molecules less soluble than smaller ones? To give a general answer to this, we need to return to our consideration of the entropy of mixing. When we have solute and solvent molecules of the same size, the entropy of mixing is maximal. As the solute molecules become larger, the number of arrangements with solvent molecules (on an equal weight basis) decreases. The wheat glutelins (glutenins) are considered to be some of the largest protein molecules in nature, reaching molecular weights in the tens of millions (Wrigley 1996).

Solubilization of gluten proteins

No solvent capable of complete solubilization of cereal proteins without alteration of their structure has been reported. The best solvents have been dilute solutions of the ionic detergents: sodium dodecyl sulfate (SDS) and cetyl trimethyl ammonium bromide (CTAB). These compounds complex with proteins to confer a negative (SDS) or positive (CTAB) electrical charge, thus compensating for the low-charged group frequency. Nevertheless, for wheat, 20–30% of the total gluten protein is usually not solubilized by these solvents. The protein that is not solubilized is predominantly the large-size glutenins.

One procedure that has been successful in achieving close to complete solubilization of gluten protein has been the application of ultrasound or sonication. It has been found that application of high stress to gluten dispersions (high-speed mixing or sonication) causes scission of the largest glutenin molecules, thus inducing solubilization (MacRitchie 1975; Singh

Table 9.2 Summarized Data for Amino Acid Composition of Protein from Different Cereals

	Wheat	Rye	Barley	Corn	Gliadin	Glutenin	Ref.
Glutamine[a]	26.7 (71)	21.4 (63)	20.6 (67)	—	—	—	d,e
+ asparagine	25.9 (65)	18.8 (54)	19.1 (56)	11.4 (45)	34.2 (92)	26.5 (82)	d,e
Proline	11.1	10.4	10.9	—	—	—	d,e
	12.1	12.0	12.0	9.8	13.8	10.3	
½ Cystine	2.6	2.3	2.5	—	—	—	d,e
	1.6	2.9	2.9	3.1	0.39	-0.34	
NPS[b]	0.34	0.34	0.36	—	—	—	d,e
	0.35	0.35	0.37	0.36	0.39	-0.34	
CGF[c]	0.16	0.19	0.17	—	—	—	d,e
	0.13	0.18	0.17	0.19	-0.08	0.13	

[a] Value in brackets is degree of amidation.
[b] NPS = trp, ile, tyr, phe, pro, leu, and val residues/total number of residues.
[c] CGF = aspartic and glutamic acids, his, lys, and arg residues/total number of residues.
[e] Ewart (1967), analyses of flours.
[d] Tkachuk and Irvine (1969), analyses of grains.

et al. 1990). When conditions are suitably chosen, sonication is reported to extract over 95% of the total wheat protein (Pasaribu, Tomlinson, and McMaster 1992). Sonication therefore changes the protein; however, the great advantage is that only the largest molecules are broken and these are split preferentially at their centers (Bueche 1960).

The fragments are therefore large molecules that fall within the glutenin size range, and the separation by size-exclusion HPLC (SE-HPLC) is unaffected if time and intensity of sonication are optimum. Thus, sonication combined with SE-HPLC provides a sound method for determining the relative proportions of the main protein classes. If the solubilized protein is to be used for measuring molecular weight distribution or functional properties, however, the changes induced by sonication must be taken into account.

The hydrophobic effect

It is well known that water is a good solvent for ions and polar molecules. On the other hand, it is a poor solvent for nonpolar molecules such as hydrocarbons. Perhaps surprisingly, the insolubility of hydrocarbons and other nonpolar compounds is not due to a positive enthalpy effect. The enthalpy of mixing hydrocarbons with water is found to be either very small or negative. Therefore, the positive value for ΔG must arise from a negative entropy change. This effect, called the hydrophobic effect, assumes importance in many processes, including solubilization and the adsorption of compounds at interfaces.

Water molecules participate in hydrogen bonding. These noncovalent bonds are continually forming and breaking. A snapshot at any instant shows a hydrogen bonded structure. When a nonpolar molecule is introduced into water, the hydrogen bonds cannot form with the nonpolar groups, so hydrogen bonds in the vicinity are broken. In order to form new hydrogen bonds, the water molecules near the nonpolar surface orient so as to re-form the hydrogen bonded structure. This produces a structuring effect in the vicinity that is believed to extend several molecular diameters into the aqueous phase, as shown in Figure 9.5. This ordering of water molecules and the accompanying negative entropy change mainly contribute to the positive free energy of mixing of water with nonpolar compounds, causing immiscibility.

Effects of neutral salts on solubility

The lyotropic or Hoffmeister series is a series of salts in which a gradation in properties including solubility is observed. This is illustrated in Table 9.3, in which the solubilities of gluten protein in 1.0 M solutions of sodium salts with different univalent anions are tabulated (Preston 1981).

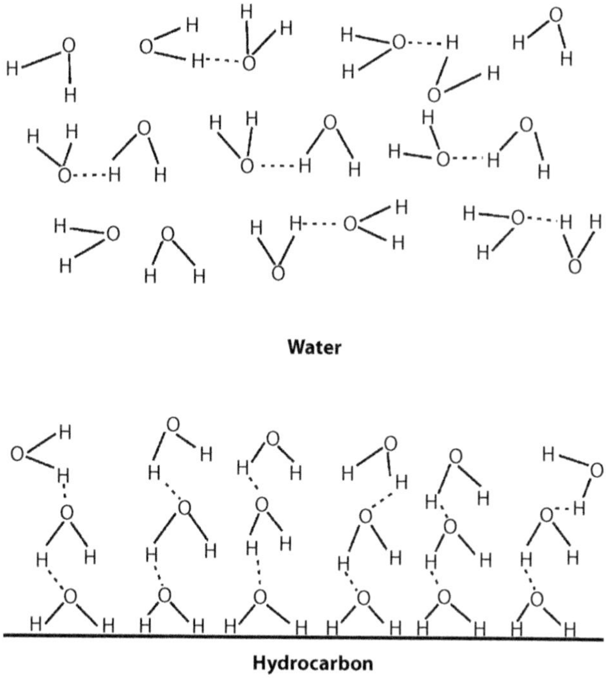

Water

Hydrocarbon

Figure 9.5 Schematic illustration of the effect of introducing a hydrophobic surface into water on the adjacent hydrogen bonding structure of the water.

Table 9.3 Effects of 1.0 M Solutions of Sodium Salts on Solubility (%) of Defatted and Nondefatted Gluten Protein

	Gluten	
Salt	Defatted	Nondefatted
None (H₂O)	25.0	29.5
NaF	Trace	Trace
NaCl	5.8	5.2
NaBrO₃	5.4	6.4
NaBr	19.9	Not determined
NaClO₄	31.3	28.7
NaI	51.7	54.6
NaSCN	61.5	59.1
HAc (0.05 M)	70.4	77.2
Lactic (0.005 M)	71.4	75.8

Source: Preston, K. R. 1981. *Cereal Chemistry* 58:317–324.

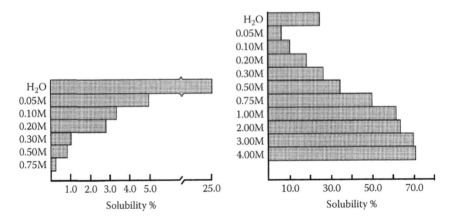

Figure 9.6 Effects of sodium fluoride (left) and sodium thiocyanate (right) on the solubility of gluten protein. (Data from Preston, K. R. 1981. *Cereal Chemistry* 58:317–324.)

As the series goes from fluoride to thiocyanate, the solubility increases dramatically. Effects of the different salts are related to their effects on the water structure surrounding the protein, especially the nonpolar groups at the periphery. The thiocyanate ion is a chaotropic agent; that is, it breaks the hydrogen bonded structure of water, increasing the entropy and contributing to higher solubility. Fluoride ion, by contrast, is a nonchaotropic agent: It enhances water structuring and consequently reduces solubility. Figure 9.6 shows the opposite effects of different concentrations of these two salts (sodium thiocyanate and sodium fluoride) on the solubility of gluten protein (Preston 1981).

Demonstrations

1. The reaction between ammonium thiocyanate and barium hydroxide octahydrate provides a good example of a highly endothermic reaction. Amounts of the two solids (stoichiometric quantities may be calculated from the equation below) are placed in a conical flask and shaken vigorously. A small amount of water is placed on a light sheet of board or Masonite (about 15 cm square) using a wash bottle. After the mixture in the flask has liquefied, the flask is placed firmly on the board so that a film of water separates the flask from the board. After a few minutes, the flask is raised. The water film is frozen, causing the flask to be strongly attached to the board.

   The reaction absorbs a large amount of heat from the surroundings, causing the temperature to fall below 0°C to freeze the water film and cement the board to the flask. The equation for the reaction is

$$Ba(OH)_2\ 8H_2O + 2\ NH_4SCN = Ba^{2+} + 2SCN^- + 2NH_3 + 10H_2O$$

2. A simple demonstration of the tendency toward greater disorder may be made by using small beads of two different colors (e.g., black and white). A layer of one color is placed on a layer of the other in a glass tube. This is the state of lowest probability. When the tube is shaken (simulating the effect of kinetic energy on molecular motion), the beads form a homogeneous mixture, thus assuming the most probable state.

Exercises

1. In demonstration 1, there must be a large increase in disorder to offset the increase in enthalpy in order for the reaction to proceed. What processes that cause the increase in disorder can be inferred from the equation for the reaction?
2. In demonstration 2, is it impossible to revert to the original ordered state of two separate layers by continuing to shake the tube?
3. Write brief notes to attempt explanations of the following observations:
   a. A protein A is insoluble in an aqueous solution. The heat of solution ΔH is negative; that is, there is an emission of heat by the protein A–water interaction.
   b. A protein B of molecular weight 15,000 is highly soluble in aqueous solution. However, when the protein is cross-linked to give an average molecular weight of 450,000, the protein precipitates from solution.
   c. A protein C had very low solubility at pH 5.5. However, it was completely soluble at pH 3.0 and pH 8.0.

References

Bigelow, C. C. 1967. On the average hydrophobicity of proteins and the relation between it and protein structure. *Journal of Theoretical Biology* 16:187–211.

Bueche, F. 1960. Mechanical degradation of high polymers. *Journal of Applied Polymer Science* 4:101–106.

Dunnill, P. 1965. How proteins acquire their structure. *Science Progress* (Oxford) 53:609–619.

Ewart, J. A. D. 1967. Amino acid analyses of cereal flour proteins. *Journal of the Science of Food and Agriculture* 18:548–552.

Fisher, H. 1964. A limiting law relating the size and shape of protein molecules to their composition. *Proceedings of the National Academy of Science U.S.A.* 51:1285–1291.

Iqbal, M., and Verrall, R. E. 1988. Implications of protein folding activity schemes for volumes and compressibilities. *Journal of Biological Chemistry* 263:4159–4165.

MacRitchie, F. 1975. Mechanical degradation of gluten proteins during high speed mixing of doughs. *Journal of Polymer Science Polymer Symposium* 49:85–90.

———. 1990. *Chemistry at interfaces.* San Diego, CA: Academic Press.

Osborne, T. B. 1907. The proteins of the wheat kernel. Washington, D.C.: Carnegie Institute, publication no. 84.

Pasaribu, S. J., J. D. Tomlinson, and G. J. McMaster. 1992. Fractionation of wheat flour proteins by size exclusion—HPLC on an agarose-based matrix. *Journal of Cereal Science* 15:121–136.

Preston, K. R. 1981. Effects of neutral salts upon wheat gluten protein properties. I. Relationships between the hydrophobic properties of gluten proteins and their extractability and turbidity in neutral salts. *Cereal Chemistry* 58:317–324.

Singh, N. K., G. R. Donovan, I. L. Batey, and F. MacRitchie. 1990. Use of sonication and size-exclusion HPLC in the study of wheat flour proteins. I. Dissolution of total proteins in unreduced form. *Cereal Chemistry* 67:150–161.

Tkachuk, R., and G. N. Irvine. 1969. Amino acid composition of cereals and oilseeds whole meals. *Cereal Chemistry* 46:206–218.

Wrigley, C. W. 1996. Giant proteins with flour power. *Nature* 381:738–739.

chapter ten

Characterization of cereal proteins

Introduction

The discussion of protein solubility in Chapter 9 leads logically into the next topic because it is usually necessary to have proteins in solution in order to characterize them. Of course, some important characterization can be made by hydrolyzing the proteins (e.g., determination of the amino acid composition and sequence). What we are mainly concerned with in this chapter is the measurement of molecular composition and properties of proteins or their subunits in or close to their native state. This is required if our aim is to relate composition to functional properties. The classification of cereal proteins into four groups based on solubility was described in the previous chapter. This classification still proves to be valuable for researchers. However, in recent times, techniques have been developed that enable more accurate quantification of cereal protein composition.

Size-exclusion high-performance liquid chromatography

High-performance liquid chromatography (HPLC) was first applied to cereal proteins by Bietz (1986). Size-exclusion HPLC (SE-HPLC) has been used to quantify cereal proteins; it separates proteins according to molecular size. Wheat proteins have been the most studied of the cereals. The preferred solvent has been dilute sodium dodecyl sulfate (SDS)/buffer. A wheat flour suspension (~1.0 mg/mL) is sonicated under conditions (time, intensity) that solubilize almost all, if not all, of the protein without altering quantitation of the main protein classes (see discussion in Chapter 9).

Although the molecular weight distribution (MWD) is changed by sonication, the degradation products of the polymeric proteins are large and elute in the first peak of the chromatogram. Separation into the three main protein groups—polymeric proteins, gliadins, and albumin/globulins—is obtained, as seen in Figure 10.1, using a Biosep-SEC-S4000 column. With the columns presently available, only a partial resolution of the polymeric protein is achieved. The fraction of polymeric protein of highest molecular weight elutes in the void volume; that is, there is no retardation of this fraction as it travels through the column.

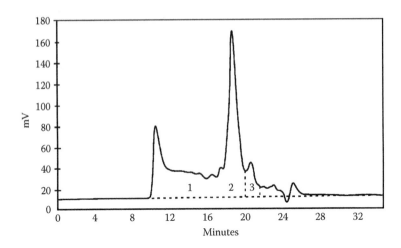

Figure 10.1 SE-HPLC of wheat flour proteins. Peak 1: polymeric proteins; peak 2: gliadins; peak 3: albumins/globulins.

Separation of wheat proteins

A simple classification of wheat proteins is that based on the chromatogram of Figure 10.1. The three main groups include the following:

1. *Polymeric proteins* are multichain proteins in which the chains (subunits) are linked by disulfide bonds. They are composed mainly of glutenins with small amounts of high molecular weight albumins, which are mainly beta-amylases (Gupta, Shepherd, and MacRitchie 1991) and high molecular weight globulins or triticins (Singh and Shepherd 1985). The glutenins make up roughly half the gluten protein. The gluten proteins are responsible for the viscoelastic properties of dough and make up roughly 85% of the total wheat endosperm protein.
2. *Gliadins* belong to the first group of monomeric proteins and make up the other half of the gluten proteins. They are single chain proteins whose amino acid composition is closer to the glutenins than to the other monomeric proteins, the albumins and globulins. The omega-gliadins are eluted in one or two small peaks at earlier elution times than the main gliadin peak containing the alpha-, beta-, and gamma-gliadins. Omega-gliadins have molecular weights in the order of 70,000, whereas the molecular weights of the other gliadins fall mainly in the range of 30,000–40,000.
3. *Monomeric albumins and globulins* are proteins that have molecular weights in the range of 20,000–30,000 and elute in a small peak at elution times slightly greater than the gliadins.

Composition and properties

The similar amino acid composition of the two groups of gluten proteins may reflect a common genetic background. Their essential difference is in molecular size. Because of their polymeric nature, glutenins have molecular weights estimated to be in the millions and probably tens of millions. Although the three main groups of proteins are well defined by SE-HPLC (Figure 10.1), other techniques are required to separate and quantify individual proteins and polypeptides (subunits).

The main techniques that can do this are reversed-phase (RP)-HPLC and several electrophoretic methods. The term "reversed phase" arose when the conventional hydrophilic columns were coated with a hydrocarbon (C8 or C18). Separation by these columns is based on differences in hydrophobicity. Some of the electrophoretic methods useful for studying cereal proteins are discussed in the next section.

Electrophoresis of gluten proteins

One of the most suitable procedures for separation of gliadin components has been electrophoresis in an acid medium (A-PAGE or acid-polyacrylamide gel electrophoresis). This has been used to separate the mixture of gliadin components into four major groups in order of increasing molecular size and therefore decreasing mobility: alpha-, beta-, gamma-, and omega-gliadins (Bushuk and Sapirstein 1991). The classification of gliadins into their four groups is shown in Figure 10.2 together with the three reference bands suggested by Bushuk and Sapirstein for calibration.

The chromatogram of Figure 10.1 enables the polymeric proteins to be separated from the gliadins and other monomeric proteins. However, in order to determine the composition of the proteins in terms of their individual proteins or subunits, it is necessary to split the interchain disulfide bonds by a reducing agent such as 2-mercaptoethanol or dithiothreitol. The preferred technique for characterizing the reduced protein is sodium dodecyl sulfate-PAGE (SDS-PAGE). The medium here is a solution of SDS where the proteins and subunits complex with SDS and acquire negative charges.

Unlike A-PAGE, where separation is based on electrical charge as well as molecular size, SDS-PAGE separates almost exclusively on the basis of molecular size. Glutenins are large polymers formed from two main types of subunits. These are named high molecular weight (or A) glutenin subunits (HMW-GS) and low molecular weight glutenin subunits (LMW-GS). The HMW-GS have MWs in the range of 80,000–120,000 as measured by SDS-PAGE. The LMW-GS are divided into two groups: the B subunits (MW = 40,000–55,000) and C subunits (MW = 30,000–40,000). When the total flour protein is reduced and run on SDS-PAGE, the HMW-GS are clearly separated and give bands with low mobility near the top of the gel.

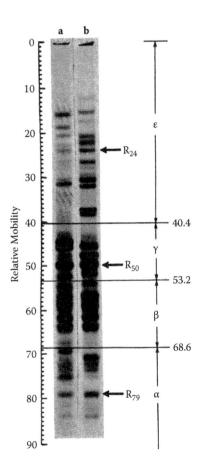

Figure 10.2 A-PAGE of gliadins showing the four main classes—alpha, beta, gamma, and omega—and the three bands used for calibration. (Reproduced with permission from Bushuk, W., and H. D. Sapirstein. 1991. In *Gluten proteins 1990*, ed. W. Bushuk and T. Tkackuk, 454–458. St. Paul, MN: American Association of Cereal Chemists.)

However, because many of the other proteins and subunits have similar sizes, these bands overlap in the gel.

Several approaches have been devised to separate the different groups in the complex mixture and to permit individual proteins or subunits to be identified and quantified. A two-step SDS-PAGE procedure was introduced by Singh and Shepherd (1988) in which the unreduced total protein is first loaded on the gel. The gliadins and other monomeric components are allowed to run ahead of the polymeric components, which form a smear near the origin. The proteins from the smear are then removed and run under reducing conditions on SDS-PAGE. The result of this second step for a

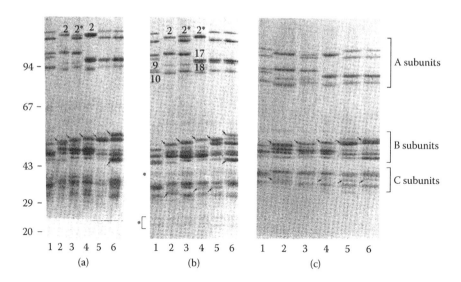

Figure 10.3 One-dimensional SDS-PAGE separations of glutenin subunits from six wheat cultivars. The one-step procedure (a) with and (b) without alkylation; (c) the second step of the two-step method. The positions of MW markers are shown on the left side of the figure. (Gupta, R. B., and F. MacRitchie, F. 1991. *Journal of Cereal Science* 14:105–109.)

set of six cultivars is shown in Figure 10.3c. This shows the slowest mobility bands (*A* subunits or HMW-GS) and, below them, two sets of faster mobility bands—the *B* and *C* subunits that, together, comprise the LMW-GS.

Another procedure (one-step SDS-PAGE) replaces the first step of the two-step SDS-PAGE procedure by solubilization and removal of the monomeric proteins. Different solvents have been used to effect the solubilization (Gupta and MacRitchie 1991; Singh, Shepherd, and Cornish 1991). Results of this one-step SDS-PAGE are shown in Figures 10.3a and 10.3b. One of the problems with this method (and also with the two-step procedure) is that a sharp separation of monomeric from polymeric proteins is not achieved; that is, some polymeric proteins (those of lower molecular weight) are solubilized with the monomeric proteins and some monomeric proteins may remain in the residue. The lack of a sharp separation can cause an appreciable error if the techniques are used for quantifying the proportions of the different glutenin subunits.

Procedures for effecting sharper separation of monomeric from polymeric fractions based on solubility have been reported by Fu and Kovacs (1999) and Fu and Sapirstein (1996). The extent of the error has been investigated by Cinco-Moroyoqui (2001), who introduced an alternative procedure for quantifying glutenin subunits. This consisted of fraction collection of the polymeric protein from SE-HPLC followed by reduction and quantification of the subunits by another technique, such as RP-HPLC. This procedure considerably reduces errors encountered in fractionation by solubilization due to overlap of different solubility fractions (Cinco-Moroyoqui and MacRitchie 2008).

Other electrophoretic methods

Many other electrophoretic techniques can be valuable for characterizing cereal proteins in addition to the one-dimensional A-PAGE and SDS-PAGE methods described before. The use of gradient SDS-PAGE was mentioned in Chapter 4 in connection with enhancing the resolution of proteins associated with grain hardness (Greenwell and Schofield 1986). Two-dimensional diagonal electrophoresis (unreduced first dimension, reduced second dimension) is able to distinguish whether proteins are single chain or disulfide bonded (Gupta et al. 1991). Isoelectric focusing uses a gel with a gradient of pH. When protein molecules arrive at a region of the gel where the pH coincides with their isoelectric points, they become stationary. This technique is particularly suitable for separating cereal proteins, which comprise a complex mixture with each component differing slightly in isoelectric point (Wrigley and Shepherd 1973).

Another technique that has developed more recently is capillary electrophoresis. This is similar to one-dimensional electrophoresis but utilizes a thin capillary filled with electrolyte. It separates on the basis of electrical charge and frictional forces and has been used to characterize cereal proteins (Lookhart et al. 1996). Lab-on-a-chip is a recent innovative technology that can be used to effect chromatographic and electrophoretic separations, including capillary electrophoresis (Uthayakumaran, Batey, and Wrigley 2005). It utilizes a network of channels and wells that are etched onto glass or polymer microchips to create minilabs. Its advantages are speed of analysis and low sample and reagent consumption (picoliters and less), as well as high reproducibility due to standardization and automation.

Molecular weight distribution

Previous chapters have emphasized the important role of the MWD in determining many functional properties of cereal proteins. Measurement of MWD has met two main problems. The first one is the difficulty of solubilization, particularly of the glutelins. The second is that many of the standard methods for measurement lose their sensitivity above a certain MW. Colligative properties such as osmotic pressure and freezing point depression are related to the number of molecules and therefore give a measure of the number-average MW (M_n).

Other measurements, such as ultracentrifugation and conventional light scattering, give a higher weighting to the larger molecules in the sample and thus give measures of the weight-average MW (M_w). All these methods are useful for smaller proteins (<100,000) but break down for the larger sized proteins. To overcome this problem, several methods that are not limited by molecular size have been introduced and will be discussed briefly.

Unextractable polymeric protein or glutenin macropolymer

As we have seen in Chapter 9, an inverse relationship exists between molecular weight and solubility. This is made use of by measuring the proportion of protein that is insoluble in a dilute SDS solution in the absence of sonication (Gupta, Khan, and MacRitchie 1993; Weegels, Hamer, and Schofield 1997). Figure 10.4 illustrates the principle of the method, using SE-HPLC to quantify the percentage of polymeric protein not solubilized in SDS-buffer without sonication. Modifications to the original procedure have been made by Bean et al. (1998) and Sapirstein and Johnson (1996).

The unextractable polymeric protein (UPP) and glutenin macropolymer (GMP) essentially measure similar parameters—the most insoluble and therefore the highest molecular weight fraction of polymeric protein. Although this is not an absolute method for molecular weight measurement, it has proved to be valuable for measuring relative molecular weight distributions. There will always be a distribution of polymeric proteins between soluble and insoluble fractions, but this distribution will heavily favor larger proteins in the insoluble residue. It seems probable that these parameters give a rough measure of the fraction of polymeric protein with MW above a certain value and the MWD of this fraction.

Multistacking SDS-PAGE

The multistacking procedure uses one-dimensional SDS-PAGE but introduces several gel layers of successively increasing density (from the top) and therefore decreasing pore size. In the original paper by Khan and Huckle (1992), five stacking gels were prepared with 4, 6, 8, 10, and 12% acrylamide concentrations and 0.6% bisacrylamide. Each stacking gel was approximately 0.5 cm in height with a resolving gel of 14% acrylamide and 0.28% bisacrylamide of 12 cm height.

The procedure clearly demonstrates the heterogeneous nature of glutenin. As the gel density increases from top to bottom, different molecular weight fractions are retained by successive layers, which can be visualized and quantified by scanning. The composition of the different fractions in terms of subunits can be measured by cutting out the gel layers and analyzing them on SDS-PAGE after reduction with an agent such as 2-mercaptoethanol.

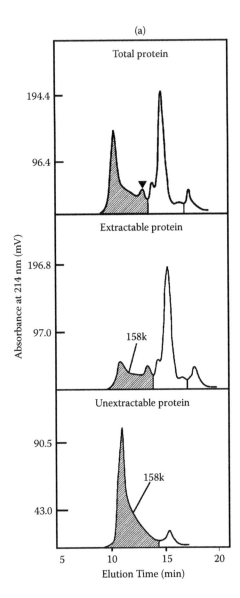

Figure 10.4 SE-HPLC of total protein (upper), SDS-buffer extractable protein (middle), and SDS-buffer unextractable protein (lower). ▼ in (a) is an omega-gliadin subpeak that coelutes with extractable polymeric proteins in the later part of the polymeric protein peak. (Reproduced with permission from Gupta, R. B. et al. 1993. *Journal of Cereal Science* 18:23–41.)

Multiangle laser light scattering

Classical light scattering measures the amount of light scattered by a solution at some angle relative to the incident beam. Theory shows that the intensity of scattered light is proportional to the product of protein concentration (milligrams per milliliter) and molecular mass. This method gives reliable measures of molecular weights for globular proteins of small size (<500,000). Cereal proteins reach higher MW and tend to be unfolded. For such molecules, light scattering varies considerably with angle. By measuring the scattering of a laser beam at a number of angles (multiangle laser light scattering; MALLS), it is possible to obtain absolute values for MW as well as the root mean square radius, and a measure of molecular conformation can be deduced.

Stringent precautions need to be taken to ensure that solutions are free of adventitious contaminants that can contribute to scattering. In combination with SE-HPLC, this method shows promise for measurement of true MWD of cereal proteins. Some of the basic precautions to be taken in using size-exclusion chromatography (SEC)-MALLS have been described by Bean and Lookhart (2001). The technique has been used by Carceller and Aussenac (2001) to obtain measures of polydispersity of glutenins and by Mendichi, Fisichella, and Savarino (2008) to obtain data on protein from different wheat cultivars.

Field flow fractionation

Another technique suited to MW measurement of large molecules is field flow fractionation (FFF), in which separation occurs in a thin channel (10–250 μm wide, approximately 30 cm long). A field is applied perpendicularly to the channel flow. The field may be centrifugal, thermal, electrical, or solvent flow. For cereal proteins, flow FFF has been used. Solvent flows through the channel and a cross-flow of solvent is introduced. Because of the thinness of the channel, a parabolic flow pattern is set up. The cross-flow causes the molecules (or particles) to be transported toward the channel wall. Smaller molecules with greater diffusivities remain near the center of the channel in the higher velocity flow. They therefore elute first—opposite to the case for SE-HPLC.

In studies of wheat proteins, Stevenson and Preston (1996) used symmetrical flow having upper and lower porous sintered glass channel walls. Wahlund et al. (1996) and Arfvidsson, Wahlund, and Eliasson (2004) used asymmetrical flow, which has only a lower porous sintered glass channel wall. The lower channel in each case is overlaid with an ultrafiltration-type membrane that allows passage of solvent molecules only.

The intrinsic diffusion coefficient of a macromolecule depends on its effective hydrodynamic diameter. The latter is influenced by the shape as well as the molecular mass (M). The hydrodynamic diameter (d) of a polymer is given by

$$d = AM^b \tag{10.1}$$

where A and b are constants that depend on the polymer-solvent system and the shape of the macromolecule.

Wahlund et al. (1996) estimated upper and lower limits for the molecular mass of glutenin fractions. The lower limit was defined as D = $0.0542M^{0.498}$ for a flexible random coil polymer and D = $0.159M^{1/3}$ for a spherical shape. Values for the upper limit were in the range of 440,000 to 11 million. In order to obtain more precise measurements, the shape of the molecule needs to be determined.

References

Arfvidsson, C., K. G. Wahlund, and A. C. Eliasson. 2004. Direct molecular weight determination in the evaluation of dissolution methods for unreduced glutenin. *Journal of Cereal Science* 39:1–8.

Bean, S. R., and G. L. Lookhart. 2001. Factors influencing the characterization of gluten proteins by size-exclusion chromatography and multiangle laser light scattering. *Cereal Chemistry* 78:608–618.

Bean, S. R., R. K. Lyne, K. A. Tilley, O. K. Chung, and G. L. Lookhart. 1998. A rapid method for quantitation of insoluble polymeric proteins in flour. *Cereal Chemistry* 75:374–379.

Bietz, J. A. 1986. High-performance liquid chromatography of cereal proteins. In *Advances in cereal science and technology,* vol. 8, ed. Y. Pomeranz. St. Paul, MN: American Association of Cereal Chemists.

Bushuk, W., and H. D. Sapirstein. 1991. Modified nomenclature for gliadins. In *Gluten proteins 1990,* ed. W. Bushuk and T. Tkackuk, 454–458. St. Paul, MN: American Association of Cereal Chemists.

Carceller, J. L., and T. Aussenac. 2001. Size characterization of glutenin polymers by HPSEC-MALLS. *Journal of Cereal Science* 33:131–142.

Cinco-Moroyoqui, F. J. 2001. Methodology for determining glutenin subunit composition of isogenic wheat lines varying in the number of high molecular weight glutenin subunits. Ph.D. thesis, Kansas State University, Manhattan, KS.

Cinco-Moroyoqui, F. J., and F. MacRitchie. 2008. Quantitation of LMW-GS to HMW-GS ratio in wheat flours. *Cereal Chemistry* 85:824–829.

Fu, B. X., and M. I. P. Kovacs. 1999. Rapid single-step procedure for isolating total glutenin proteins of wheat flour. *Journal of Cereal Science* 29:113–116.

Fu, B. X., and H. D. Sapirstein. 1996. Procedure for isolating monomeric proteins and polymeric glutenin of wheat flour. *Cereal Chemistry* 73:143–152.

Greenwell, P., and J. D. Schofield. 1986. What makes hard wheats soft? *FMBRA Bulletin* 4:3–18.

Gupta, R. B., K. Khan, and F. MacRitchie. 1993. Biochemical basis of flour properties in bread wheats. I. Effects of variation in the quantity and size distribution of polymeric proteins. *Journal of Cereal Science* 18:23–41.

Gupta, R. B., and F. MacRitchie. 1991. A rapid one-step one-dimensional SDS-PAGE procedure for analysis of subunit composition of glutenin in wheat. *Journal of Cereal Science* 14:105–109.

Gupta, R. B., K. W. Shepherd, and F. MacRitchie. 1991. Genetic and biochemical properties of some HMW albumins associated with glutenin in bread wheat. *Journal of Cereal Science* 13:221–235.

Khan, K., and L. Huckle. 1992. Use of multistacking gels in sodium dodecyl sulfate-polyacrylamide gel electrophoresis to reveal polydispersity, aggregation, and disaggregation of the glutenin protein fraction. *Cereal Chemistry* 69:686–688.

Lookhart, G. L., S. R. Bean, R. Graybosch, O. K. Chung, B. Morena-Sevilla, and S. Baenziger. 1996. Identification by high-performance capillary electrophoresis of wheat lines containing the 1AL.1RS translocation. *Cereal Chemistry* 73:547–550.

Mendichi, R., S. Fisichella, and A. Savarino. 2008. Molecular weight, size distribution and conformation of glutenin from different wheat cultivars by SEC-MALLS. *Journal of Cereal Science* 48:486–493.

Sapirstein, H. D., and W. J. Johnson. 1996. Spectrophotometric method for measuring functional glutenin and rapid screening of wheat quality. In *Gluten 96. Proceedings of the International Workshop on Gluten Proteins*, ed. C. W. Wrigley, 494–497. Melbourne: Royal Australian Chemical Institute.

Singh, N. K., and K. W. Shepherd. 1985. The structure and genetic control of a new class of disulfide linked proteins in wheat. *Theoretical and Applied Genetics* 7:79–92.

———. 1988. Linkage mapping of the genes controlling endosperm proteins in wheat. I. Genes on the short arms of group 1 chromosomes. *Theoretical and Applied Genetics* 75:628–641.

Singh, N. K., K. W. Shepherd, and G. B. Cornish. 1991. A simplified SDS-PAGE procedure for separating LMW subunits of glutenin. *Journal of Cereal Science* 14:203–208.

Stevenson, S. G., and K. R. Preston. 1996. Flow field-flow fractionation of wheat proteins. *Journal of Cereal Science* 23:121–131.

Uthayakumaran, S., I. L. Batey, and C. W. Wrigley. 2005. On-the-spot identification of grain variety and wheat-quality type by lab-on-a-chip capillary electrophoresis. *Journal of Cereal Science* 41:371–374.

Wahlund, K. G., M. Gustavsson, F. MacRitchie, T. Nylander, and L. Wannerberger. 1996. Size characterization of wheat proteins, particularly glutenins, by asymmetrical flow field-flow fractionation. *Journal of Cereal Science* 23:113–119.

Weegels, P. L., R. J. Hamer, and J. D. Schofield. 1997. Depolymerization and repolymerization of wheat gluten during dough processing. I. Relationships between glutenin macropolymer content and quality parameters. *Journal of Cereal Science* 23:103–111.

Wrigley, C. W., and K. W. Shepherd. 1973. Electrofocusing of grain proteins from wheat genotypes. *Annals of the New York Academy of Sciences* 209:154–162.

chapter eleven

Composition–functionality relationships

Introduction

In previous chapters, the composition of the main chemical components of cereals has been discussed. Lipids were discussed in Chapter 7, starch in Chapter 8, and proteins in Chapters 9 and 10. In this chapter, we will discuss some methods that have been used to deduce relationships between composition and functionality. Several approaches have been used.

Varietal surveys

The aim in this approach is to study a set of varieties, obtain some measure of composition, and test its correlation with one or more measures of functionality. It is illustrated by the classical work of Finney and Barmore (1948) in which a set of wheat cultivars was grown at different locations, resulting in a wide range of protein content for each cultivar. The bread loaf volume in an optimized baking test—an objective measure of baking potential—was measured for all flour samples and plotted against flour protein content for each variety (Figure 11.1). Two main conclusions could be drawn from the results: (1) loaf volume increased linearly with increasing flour protein content, and (2) the slope of the lines appeared to be a characteristic for each cultivar. At low protein levels, the lines tended to converge, but, as flour protein increased, the lines diverged. The variety with the highest slope could be considered to be the variety with the highest quality and the variety with the lowest slope had the lowest quality.

Many attempts have been made to relate protein composition to functional properties based on solubility separations of protein fractions. For example, it has been thought, quite logically, that the glutenin/gliadin ratio should be related to dough strength. However, attempts to establish such a relationship have often proved to be equivocal. With the advent of size-exclusion high-performance liquid chromatography (SE-HPLC) to characterize wheat proteins, greater success has been obtained with this more accurate method for measuring protein composition. Figure 11.2 shows results for the variation of extensigraph maximum resistance (R_{max})

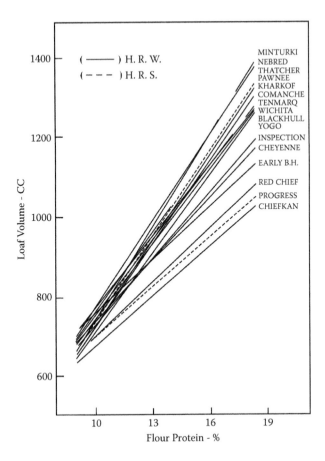

Figure 11.1 Bake-test loaf volume versus flour protein percentage for flours from a set of wheat varieties grown at different locations. H. R. W. = Hard Red Winter Wheat; H. R. S. = Hard Red Spring Wheat. (Reproduced with permission from Finney, K. F., and Barmore, M. A. 1948. *Cereal Chemistry* 25:291–312.)

with percentage of polymeric protein (mainly glutenin) in the protein of flours from a set of 15 hexaploid wheats grown at six different nitrogen fertilizer levels (Gupta, Batey, and MacRitchie 1992). A correlation coefficient (r) of 0.665*** ($n = 85$) was obtained.

The correlation is not high, but an examination of the scatter of points gives extra insight into the data. It can be seen that one variety (symbol H) has a clustering of points well above the line of best fit, while the points for another variety (symbol I) are all well below the line. It was found that the variety Halberd (H) had a much higher ratio of high molecular weight/ low molecular weight (HMW/LMW) glutenin subunits (GS) than variety Israel M68 (I). This ratio is positively correlated with MW, as has been

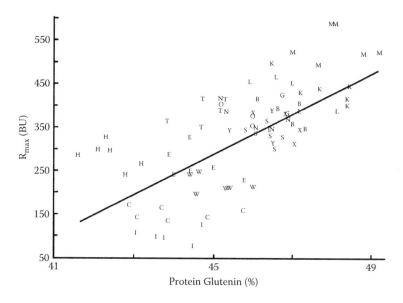

Figure 11.2 Extensigraph maximum resistance (R_{max}) versus percentage of glutenin in the protein from flour of wheat varieties grown at six different fertilizer levels. Each symbol represents a variety. The line is the line of best fit through the points. (Reproduced with permission from Gupta, R. B. et al. 1992. *Cereal Chemistry* 69:125–131.)

found from other studies (Larroque et al. 1997). The results thus suggest that two parameters determine dough strength: the percentage of polymeric protein (or the glutenin/gliadin ratio) and the molecular weight distribution (MWD) of the polymeric fraction.

Separation, fractionation, and reconstitution

A direct approach to determining composition–functionality relationships begins by separating the different chemical components. Once the components have been isolated—say, from a cereal flour—it is possible to proceed in two main ways. The first is to add the component in different amounts to the flour and measure the response in terms of some parameter of functionality. This approach has been used for elucidating the effects of lipids and proteins of wheat flour.

An example is illustrated in Figure 11.3 (cf. Figure 7.10 in Chapter 7) for the effect of varying the natural lipid level of a wheat flour on bread loaf volume in an optimized lean formula baking test. The behavior is unusual. The nonstarch flour lipid has been solvent-extracted using chloroform and added back incrementally to the defatted flour. Small additions of lipid decrease loaf volume until a minimum is reached at a lipid content intermediate between the defatted and whole flour; after this, the loaf volume increases, approaching a plateau value at lipid contents above that of the whole flour.

An important requirement in this reconstitution work is that functionality of the components does not change as a result of the separation techniques. To check this for the data shown in Figure 11.3, the results were repeated using mixtures of defatted and whole flours to give the different lipid levels, thus avoiding the effect of the extraction step on lipid functionality. The curve obtained coincided closely with that found by adding the lipid fraction, tending to confirm that it was not caused by artifacts resulting from the lipid extraction procedure (MacRitchie and Gras 1973). Similar experiments have been carried out in which the protein content of flours has been varied by addition of gluten protein. Increasing flour protein level increased loaf volume in an optimized baking test in a linear manner. This is consistent with the results shown in Figure 11.1 in which the protein contents of varieties were varied by changing the growing location (MacRitchie 1978).

The second approach that can be used is to interchange corresponding components between different flours. An example is illustrated in Table 11.1. Flours from two wheat varieties, one of better quality (A) than the other (B), were each separated into three fractions: gluten, starch, and solubles. The fractions of each flour were interchanged and bake-test loaf

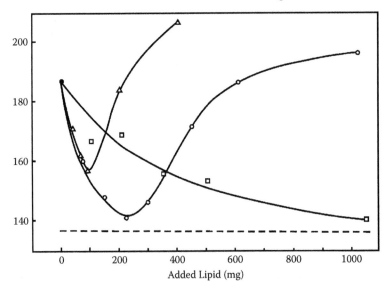

Figure 11.3 Loaf volume versus lipid added to a defatted flour. O = whole flour lipid; Δ = polar lipid; □ = nonpolar lipid; dashed line is the volume reached at the end of proof. (MacRitchie, F., and P. W. Gras. 1973. *Cereal Chemistry* 50:292–302.)

Table 11.1 Effect on Loaf Volume of Interchanging Fractions between Two Flours, A and B, of Differing Baking Performance

Gluten	Starch	Solubles	Loaf volume (cm^3)
A	A	A	182
B	B	B	161
A	A	B	182
A	B	B	180
A	B	A	183
B	B	A	160
B	A	B	163
B	A	A	163

Source: MacRitchie, F. 1978. *International Journal of Food Science Technology* 13:187–194.

Note: sd = 4 cm^3.

volumes measured in order to deduce which component contributed most to the difference in quality (MacRitchie 1978). In this type of interchange experiment, the total number of combinations of components is given by the following simple equation:

$$\text{Number of combinations} = x^n \qquad (11.1)$$

where x = number of flours and n = number of fractions.

For the two flours, each separated into three fractions, the number of possible combinations is $2^3 = 8$. The eight combinations are shown in Table 11.1. It should be noted that the first two combinations are simply the three components of each flour reassembled. It is vital that the functional properties found for the reconstituted flours be the same as those of the original flours. Only if this can be established can the results for the interchanges be considered to be reliable. An inspection of Table 11.1 shows that the gluten protein component is responsible for the differences in loaf volume.

Use of near-isogenic lines

Near-isogenic wheat lines are valuable for exploring composition–functionality relationships. These are lines that differ by only one or a few genes. They have the advantage that comparisons can be made between lines with the same genetic background—unlike varietal studies in which many genetic variations may be superimposed. Before proceeding, it will be useful to consider the genes that code for the gluten proteins— the wheat proteins that contribute mainly to functionality. Hexaploid or

common wheats, which are used for bread-making, have three complete
sets of chromosomes or genomes, each with seven pairs of chromosomes
(i.e., a total of 42 chromosomes). Proteins coded by genes are on each of
the seven groups of chromosomes but the genes controlling the gluten
proteins are located only on chromosome groups 1 and 6.

Figure 11.4 shows a schematic and simplified representation of the
chromosomal locations of these genes. Glutenin subunits are coded by
loci on group 1 chromosomes, the HMW-GS coded at the *Glu-1* loci on
the long arms, and the LMW-GS at the *Glu-3* loci on the short arms. Two
HMW-GS, denoted as x (lower mobility) and y (higher mobility) subunits,
may be expressed at each of the three *Glu-1* loci. In practice, both subunits
are expressed at *Glu-D1*, one or two at *Glu-B1*, and one or zero at *Glu-A1*

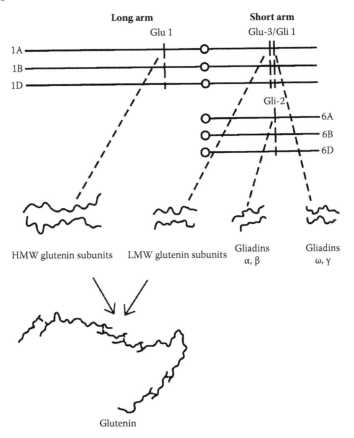

Figure 11.4 Schematic representation of the chromosomal location of genes coding for the gluten proteins. Glutenin polymerizes by a post-translational process to form the large glutenin polymers. (MacRitchie, F. 1999. *Cereal Foods World*
44:188–193.)

for common cultivars. Gamma- and omega-gliadins are coded at the *Gli-1* loci on the short arms of group 1 chromosomes. The genes coding for these gliadins are tightly linked to those coding for the LMW-GS at *Glu-3* loci. Other gliadins (alpha and beta), now considered as one group, are coded at the *Gli-2* loci on the short arms of group 6 chromosomes.

It should be noted that other minor loci for glutenin subunits have been identified on these chromosomes (Payne, Holt, and Lister 1988; Ruiz and Carrillo 1993; Pogna et al. 1995) as well as on other chromosomes (Sreeramulu and Singh 1997). It is important to realize that functional properties such as dough viscoelasticity would not be attained if only these gene products (gliadins and glutenin subunits) were synthesized. During kernel development, the glutenin subunits polymerize through disulfide bonds by a post-translational process to form the large glutenin molecules responsible for viscoelastic dough properties.

A brief summary will now be given of how near-isogenic lines have been used to gather information about composition–functionality. Before considering the three main loci coding for the gluten proteins (*Glu-1, Gli-1/Glu-3,* and *Gli-2*), it is fitting to mention the pioneering work of Sears (1954, 1966), who laid the foundations for this work by creating a series of near-isogenic lines in the wheat variety Chinese Spring. These lines included ones that were monosomic (one chromosome missing), nullisomic-tetrasomic (missing one chromosome but with a double dosage of another), and ditelosomic (missing one arm of a chromosome). Lines are now available that have deletions, additions, or allelic variation (alternative forms of a gene at the same locus). These lines not only are valuable for determining composition–functionality relationships but also have the potential to be used in breeding programs to manipulate protein composition and therefore functional properties in predictable ways.

Glu-1 loci

Two sets of near-isogenic lines that vary in the number of HMW-GS have been developed by Lawrence, MacRitchie, and Wrigley (1988) and by Payne et al. (1987); they have proved to be valuable for determining the role of this group of glutenin subunits. The set of lines produced by Lawrence et al. was obtained by crossing a mutant line of the cultivar Olympic, null at the *Glu-B1* locus, with an isogenic line of the cultivar Gabo, null at the *Glu-A1* and *Glu-D1* loci. This resulted in a set of lines in which the number of HMW-GS varied from the full complement of five down to zero. The sodium dodecyl sulfate (SDS) polyacrylamide gel electrophoresis (PAGE) patterns of the different lines are shown in Figure 11.5. The effect of deletion on functional properties is illustrated by the results shown in Table 11.2.

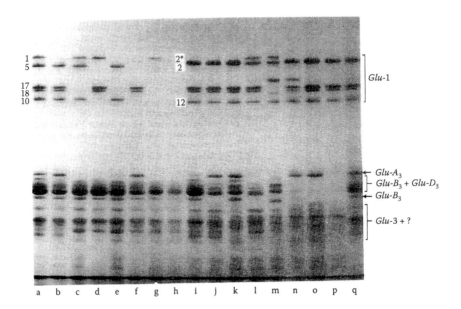

Figure 11.5 (a–h) SDS-PAGE pattern of HMW-GS set of lines and (i–q) wheat rye translocation set of lines. (Reproduced with permission from Gupta, R. B. et al. 1995. *Journal of Cereal Science* 21:103–116.)

Table 11.2 Mixograph Dough Development Time and Bake-Test Loaf Volume for Isogenic Lines Differing in HMW-GS

HMW-GS	MDDT (min)	LV (mL)
1, 17+18, 5+10	5.9	48
17+18, 5+10	4.6	46
1, 5+10	3.4	41
1, 17+18	2.7	na
17+18	3.3	46
5+10	2.4	40
1	1.7	34
Null	0.7	32

Source: Data from Lawrence, G. J. et al. 1988. *Journal of Cereal Science* 7:109–112.

Notes: MDDT = mixograph dough development time; LV = loaf volume in a microbaking test using 10-g flour.

As the number of HMW-GS is reduced from five to zero, both mixograph dough development time (MDDT) and bake-test loaf volume decrease dramatically. This result supports the work of Payne, Corfield, and Blackman (1981), which had shown, from varietal studies, that although making up only 10–15% of the total wheat protein, the HMW-GS was the protein component that contributed most to functional properties. Another interesting result for this set of HMW-GS null lines was that the total quantity of HMW-GS decreased approximately linearly with decreasing numbers of subunits (Figure 11.6). Thus, for these proteins at least, their quantity was closely related to their number and therefore the number of genes coding for them.

In addition to deletion lines, other sets of near-isogenic lines are those in which allelic variants occur. Lawrence made use of natural mutants in a number of wheat cultivars to select near-isogenic lines that differed only in HMW-GS at specific loci (Lawrence et al. 1987). Extensigraph measurements on flours from pairs of varieties differing in allelic composition gave direct support to the varietal studies by Payne and colleagues. (Payne, Corfield, and Blackman 1981). These studies showed that certain alleles coding for HMW-GS (e.g., *Glu-D1d* coding for HMW-GS 5+10) were associated with dough strength and good bread-making performance

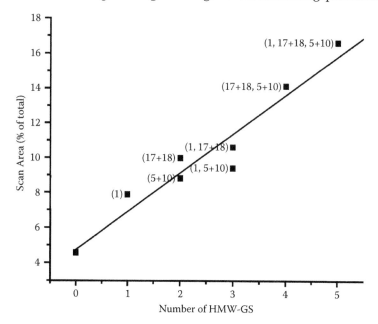

Figure 11.6 Scan area by densitometry of HMW-GS by SDS-PAGE versus number of subunits for the set of HMW-GS null lines. (From data of Lawrence, G. J. et al. 1988. *Journal of Cereal Science* 7:109–112.)

and that their allelic counterparts (e.g., *Glu-D1a* coding for HMW-GS 2+12) were associated with low dough strength and poor bread-making.

The protein composition–functionality relationships for three pairs of these cultivars were studied by Gupta and MacRitchie (1994). Each cultivar had two allelic variants at *Glu-D1*; one had the *Glu-D1a* allele (HMW-GS 2+12) and the other the *Glu-D1d* allele (HMW-GS 5+10). The results are shown in Figure 11.7. Neither the percentage of flour protein nor the percentage of polymeric protein (PPP) is seen to be correlated with the measure of dough strength (mixograph dough development time).

However, there is a good match with the percentage of unextractable polymeric protein (UPP). This is a direct confirmation of the varietal studies of Payne et al. (1981), which showed that HMW-GS 5+10 were associated with greater dough strength than HMW-GS 2+12. It also indicated that this observation could be explained by an MWD shifted to higher

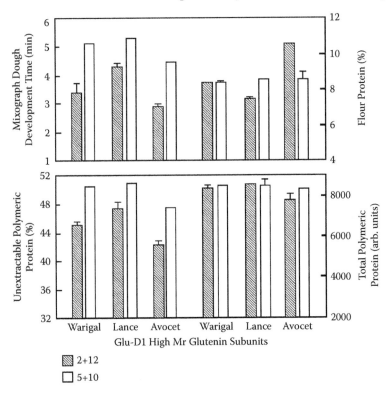

Figure 11.7 Mixograph dough development time (top left), percent of flour protein (top right), percent of polymeric protein (bottom right), and percent of unextractable polymeric protein (bottom left) for near-isogenic lines of three varieties having either *Glu-D1a* (HMW-GS 2+12) or *Glu-D1d* (HMW-GS 5+10). (Gupta, R. B., and F. MacRitchie. 1994. *Journal of Cereal Science* 19:19–29.)

Table 11.3 Protein Composition and Mixograph Dough Development Times for
Near-Isogenic Lines (NILs) of Tetraploid (Durum) Wheat Cultivar Svevo

Line	Glu-A1	Glu-B1	MDDT (min)	PPP	UPP (%)
Svevo	Null	7+8	5.1	47.2	54.1
Svevo NIL	5+10	7+8	15.0	49.8	62.1

Source: Data from Lafiandra, D. et al. 2000. In *Wheat gluten*, ed. P. R. Shewry and A. S. Tatham,
51–54. London: Royal Society of Chemistry.

Notes: MDDT = mixograph dough development time; PPP = percent polymeric protein;
UPP = unextractable polymeric protein.

molecular weights when HMW-GS 5+10 were present, based on the positive relation between MWD and UPP.

Another way of introducing variation to produce near-isogenic
lines is by translocation of loci. This approach has been utilized by
Lukaszewski using chromosomal engineering (Ammar, Lukaszewski,
and Banowetz 1997). An example is the translocation of the *Glu-D1d* allele
from a hexaploid onto chromosome 1A of a tetraploid wheat. The effect of
replacing the null *Glu-A1* allele by the *Glu-D1d* allele is seen in Table 11.3.
A dramatic increase in dough strength (measured by mixograph dough
development time) is achieved. The PPP increases some 2.6%; however,
the increase of 8.0% in the UPP may be more relevant to explaining the
increase in dough strength.

Factors influencing MWD

In view of the overriding role that MW plays in determining physical
properties, it is fitting to pause in order to summarize the compositional
variables that have been found to affect the MW of wheat polymeric protein and, therefore, the MW of all the cereal polymeric proteins. Three
main variables have been found to influence the MWD:

1. The positive relationship of the HMW/LMW-GS ratio to the MWD,
   as may be concluded from the results of Figure 11.2 (Gupta et al.
   1992). Other results, such as those of Larroque et al. (1997), support
   this conclusion.
2. The specific HMW-GS that arise from allelic variation (e.g., the allelic
   pairs 5+10 and 2+12 as discussed previously; cf. Figure 11.7).
3. The presence of proteins with a single cysteine residue that act as
   chain terminators and thus tend to shift the MWD to lower values
   (Masci et al. 1993).

Genetic transformation

Introduction of genes by transformation techniques is another approach that can give information about composition–functionality. It also has the potential to be used in breeding programs to introduce specific genes into elite varieties to modify end-use properties. Examples of this approach can be found in the recent publications of Blechl et al. (2007) and Rakszegi et al. (2008).

In Blechl and colleagues' study, transformants were produced in which the levels of one or both HMW-GS Dx5 and Dy10 were increased by increasing the copy numbers of their corresponding genes. The properties of the lines were evaluated by mixograph measurements on their flours. Increase of copy numbers produced large increases in dough mixing strength and tolerance. These results, in which HMW-GS amounts are increased above those of the parent line, therefore extend the results of Table 11.1, in which the amounts of these subunits were incrementally decreased. Increases in Dx5 subunits were found to have larger effects than comparable increases in Dy10 subunits; this supports evidence from other studies (Lafiandra et al. 1993).

Gli-1/Glu-3 loci

Lines with deletions at *Glu-3* (LMW-GS) similar to the HMW-GS null lines are difficult to obtain due to the tight linkage of these loci with loci coding for omega- and gamma-gliadins. The closest materials are the wheat/rye translocation lines developed by Gupta and Shepherd (1993). In these lines, one, two, or all three of the short arms of the group 1 chromosomes of wheat are replaced by the short arm of chromosome 1 of diploid rye. This effectively results in the deletion of *Glu-3* loci coding for LMW-GS and their replacement by the *sec-1* locus of rye, coding for secalins.

Secalins are single chain proteins similar to gliadins. The SDS-PAGE patterns for the wheat/rye translocation lines (single, double, and triple) are included in Figure 11.5 together with those for the set of HMW-GS null lines. An interesting comparison of the dough strengths of the two sets of lines (measured by extensigraph maximum resistance, R_{max}) is shown in Figure 11.8. As glutenin subunits are deleted for the two sets, R_{max} decreases. However, the decrease per unit weight is seen to be greater for deletion of the HMW-GS; this is consistent with other data suggesting that the latter glutenin subunits have the greater effect on functionality (Gupta et al. 1991).

Other lines have been developed in which there is either deletion of the *Gli-1/Glu-3* locus or allelic variation at this locus. Table 11.4 presents some data for three cultivars in which a comparison is made for MDDT when the *Gli-B1/Glu-B3* locus is present or absent (Gianibelli et al. 1998).

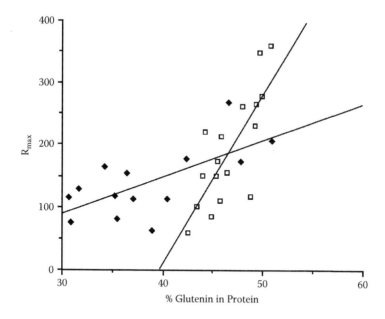

Figure 11.8 Extensigraph maximum resistance (R_{max}) versus percentage glutenin in the protein for the HMW-GS null lines (open squares) and wheat/rye translocation lines (full squares). (Reproduced with permission from Gupta, R. B. et al. 1991. In *Gluten proteins* 1990, ed. W. Bushuk and R. Tkachuk, 71–80. St. Paul, MN: American Association of Cereal Chemists.)

Table 11.4 Effect of Deletion of *Gli-B1/Glu-B3* Locus on Percent Polymeric Protein in Total Protein and Mixograph Dough Development Time

Line	Gli-B1/Glu-B3	PPP	MDDT (min)
Oderzo	+	51.1	4.9
Null	−	45.4	4.2
San Pastore	+	50.5	2.6
Null	−	42.5	2.2
Spada	+	52.2	4.5
Null	−	46.6	3.6

Source: Data from Gianibelli, M. C. et al. 1998. In *Proceedings of the 9th International Wheat Genetics Symposium,* vol. 4, ed. A. E. Slinkard. Saskatchewan, Canada: University of Saskatchewan.

Notes: MDDT = mixograph dough development time; PPP = percent polymeric protein.

Table 11.5 Protein Composition and Mixograph Dough Development Times for Flours from Lines Varying at *Gli-D1/Glu-D3*

Line	Gli-D1/ Glu-D3	PPP	UPP (%)	MDDT (min)
Oasis	CS	46.7	47.7	4.4
	CNN	42.7	51.3	5.3
Thomas	CS	54.7	49.4	4.7
	CNN	53.2	52.6	5.6

Source: Data from Gianibelli, M. C. et al. 1998. In *Proceedings of the 9th International Wheat Genetics Symposium,* vol. 4, ed. A. E. Slinkard. Saskatchewan, Canada: University of Saskatchewan.

Notes: PPP = percent polymeric protein in total protein; UPP = unextractable polymeric protein; MDDT = mixograph dough development time; CS = Chinese Spring allele; CNN = Cheyenne allele.

The effect of deletion of this locus is to decrease the PPP with a corresponding decrease in the MDDT.

Table 11.5 shows results for two cultivars in which allelic variation occurs at the *Gli-1/Glu-3* locus—in this case, for the D-genome (Gianibelli et al. 1998). In one case, the allele corresponds to that found in the good-quality variety Cheyenne (CNN); in the other, it corresponds to the allele in the poorer quality variety Chinese Spring (CS). Replacement of the CS allele by the CNN allele causes the PPP to drop, but the UPP is increased. What can be concluded from this is that the ratio of LMW-GS/gliadins is lower in the CNN line but that the glutenin in the CNN line has the MWD shifted to higher values, thus increasing the dough strength as measured by the MDDT.

Gli-2 loci

Gli-2 loci code for the alpha/beta gliadins that are now considered to be one group. It has been suggested that some LMW-GS are coded by genes on chromosome 6; however, their amounts appear to be small relative to those coded by the *Gli-1/Glu-3* loci. Table 11.6 shows some composition–functionality data for the Russian variety Saratovskaja and two near-isogenic lines in which either the *Gli-A2* or the *Gli-D2* allele is not expressed (Gianibelli et al. 1998). It might be expected that deletion of the *Gli-2* loci could be a potential approach to increasing the glutenin/gliadin ratio to correct the problem of weak dough properties. From Table 11.6, the effect of deletion in the lines studied thus far is to increase this ratio, as would be expected, but only by a relatively small amount.

Table 11.6 Effect of Deletion of *Gli-2* Locus on Percent Polymeric
Protein and Mixograph Dough Development Time

	Gli-A2	Gli-D2	PPP	MDDT (min)
Saratovskaja				
Normal	+	+	56.5	4.9
Gli-A2 null	–	+	60.5	5.7
Gli-D2 null	+	–	59.6	5.9

Source: Data from Gianibelli, M. C. et al. 1998. In *Proceedings of the 9th International Wheat Genetics Symposium,* vol. 4, ed. A. E. Slinkard. Saskatchewan, Canada: University of Saskatchewan.

Notes: PPP = percent polymeric protein in total protein; MDDT = mixograph dough development time.

Waxy lines

As has been seen, variation of protein composition accounts for many changes in functionality relevant to end uses. Variation in starch composition also assumes importance for certain applications. The term "waxy" has been used to describe cereals that have unusually high proportions of amylopectin. Waxy maize starch has been known for a long time. More recently, attention has been given to waxy wheat. Starch synthesis occurs through reaction pathways governed by different enzymes. Some of these are the granule-bound starch synthase (GBSS), responsible for amylose synthesis, and the branching

Table 11.7 Differential Effect of Wx-A1, -B1, and -D1 Proteins

Type	Wx-A1	Wx-B1	Wx-D1	App. am. (%)
Wild-type				
1	+	+	+	28
Single null				
2	–	+	+	27
3	+	–	+	26
4	+	+	–	27
Double null				
5	+	–	–	20
6	–	+	–	24
7	–	–	+	22
Waxy				
8	–	–	–	<1

Source: Data from Yamamori, M., and N. T. Quynh. 2000. *Theoretical and Applied Genetics* 100:32–38.

Notes: + = presence of Wx protein; – = absence of Wx protein. App. am (%) = apparent amylose content.

and debranching enzymes involved in amylopectin formation. Three related GBSS enzymes (isoenzymes) are coded by loci (denoted by *Wx-A1*, *Wx-B1*, and *Wx-D1*) located on chromosomes 7AS, 4AL, and 7DS of hexaploid wheats (S and L here denote the short and long arms of the chromosomes).

Mutant lines with single, double, and triple deletions of the GBSS enzymes have been produced. These mutants are depicted in Table 11.7 together with corresponding percentages of amylose present in the wheat lines. As the table shows, a single deletion reduces the amylose content slightly and a double deletion somewhat more; however, the triple null reduces it to practically zero. The partially waxy (single and double deletions) and completely waxy (triple deletion) lines thus provide materials with a range of starch properties and the utilization of these novel starches is currently being explored.

Relating genes to phenotypic traits

We have been looking at how certain processing parameters depend on the allelic composition of wheat. This knowledge helps us to formulate strategies for tailoring these properties to end-use requirements in breeding improved varieties. This is only a part of the more general task of relating genes to specific traits. Because of the huge gene pools present in cereals, this is being tackled by large teams of scientists.

To give one example, nested association mapping (NAM) aims to dissect the genetic basis of quantitative traits in maize, the world's number-one crop (Pennisi 2008). The long-term objective is to enhance genetic diversity so as to increase yield and nutritional value and to improve the resistance to disease, thus bridging the gap between genomics and plant breeding.

When embarking on this work, it is sobering to realize the sizes of the genetic structures in cereals. Bread wheat (*Triticum aestivum*), as we have seen, has three genomes of related progenitor species. The genetic material contains a total of 17 billion base pairs (i.e., 17 Gb), making it five times larger than the human genome. The largest wheat chromosome (3B) is more than twice the size of the entire 370 Mb rice genome. Recently, a team of Institut National de la Recherche Agronomique (INRA [National Institute for Agricultural Research]) researchers from Clermont-Ferrand, Toulouse, and Versailles, led by C. Feuillet (Paux et al. 2008), has completed a physical map of this wheat 3B chromosome; this is the first step to sequencing the wheat genome.

Use of environment

Functional properties of wheat are, of course, determined very much by genetic composition. However, environmental effects are invariably superimposed, often leading to functionality that would not be expected based

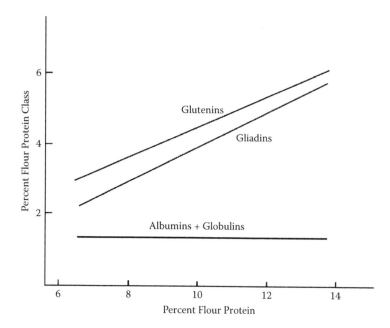

Figure 11.9 Percentage of the three main flour protein classes as a function of percentage of flour protein.

on genotype alone. This can be a challenge for breeders who must try to develop varieties that are tolerant to environmental influence. Nevertheless, the effect of environment can be used to elucidate composition–functionality relationships. This can be done if the genetic composition is kept constant and the effect of the environmental variable is measured.

Two examples of how this can be done will be described, using sulfur availability and heat stress as the two variables. Before doing this, let us consider how an important environmental variable—nitrogen availability—affects wheat protein composition. We have already seen how this variable influences flour protein content and, as a result, loaf volume in a bake test (Figure 11.1). Figure 11.9 shows how the three main wheat protein fractions vary with flour protein content. The gluten proteins increase steeply while the albumin/globulin proteins increase only slightly with increasing flour protein. Also, the slope for gliadin proteins is greater than that for glutenins. This may be why dough properties sometimes become weaker as the flour protein content attains high values.

Sulfur availability

Wheat proteins can be classified based on their sulfur content (Wrigley et al. 1984). Sulfur-rich proteins comprise LMW-GS; alpha-, beta-, and

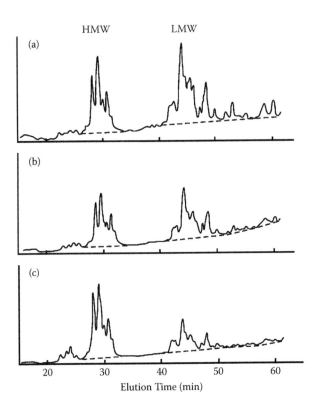

Figure 11.10 Reverse phase HPLC of reduced glutenin from Olympic flour of varying sulfur (S) and protein (P) levels. a: S = 0.146%, P = 10.4%; b: S = 0.100%, P = 7.8%; c: S = 0.075%, P = 9.7%. (MacRitchie, F., and R. B. Gupta. 1993. *Australian Journal of Agricultural Research* 44:1767–1774.)

gamma-gliadins; and LMW albumins and globulins. Sulfur-poor proteins, then, comprise HMW-GS and omega-gliadins. When sulfur availability drops below a threshold value, a shift occurs in the quantities of the different proteins. Sulfur-poor proteins increase at the expense of the sulfur-rich proteins. This effect is illustrated in Figure 11.10 for 3 samples from a set of 24 flour samples of one variety (Olympic) grown at various sulfur fertilizer levels (MacRitchie and Gupta 1993).

The figure shows how the ratio of HMW-GS to LMW-GS increases as sulfur percentage in the flour decreases. The threshold value for sulfur content appears to be about 0.15%. Figure 11.11 shows a plot of the percentage of UPP as a function of the HMW/LMW-GS ratio for the set of samples of the variety Olympic. A linear correlation coefficient of 0.699*** was obtained. This is consistent with other observations that an increase in the ratio of HMW/LMW-GS causes a shift of the MWD of the gluten protein to higher values. The effect on functionality is to shift

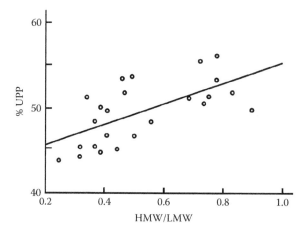

Figure 11.11 Unextractable polymeric protein (percent) versus HMW/LMW-GS ratio measured from Olympic flours of varying sulfur contents. (MacRitchie, F., and R. B. Gupta. 1993. *Australian Journal of Agricultural Research* 44:1767–1774.)

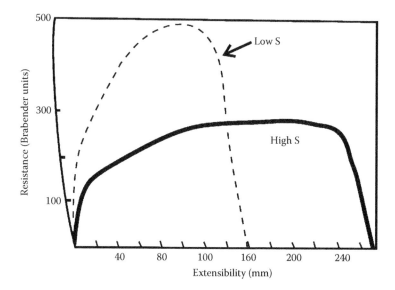

Figure 11.12 Extensigrams of flours from high and low sulfur contents. (Reproduced with permission from Wrigley, C. W. et al. 1984. *Journal of Cereal Science* 2:15–24.)

Table 11.8 Data for Near-Isogenic Lines of Lance for Different Temperature Regimes. Lance C has *Glu-D1d* allele, Lance A has *Glu-D1a* allele.

Lance line	Temperature regime	2001 Time % UPP starts steep increase (DAA)	% UPP at maturity[a]	SDS-sed. volume[b] (mL)	2002 Time % UPP starts steep increase (DDA)	% UPP at maturity[a]
C	1	25	51	4.7	25	48
A	1	31	43	4.1	28	42
C	2	25	47	3.0	22	47
A	2	28	37	1.2	25	41
C	3	19	45	1.4	19	45
A	3	22	34	1.0	25	33

Source: Reproduced with permission from Irmak, S. et al. 2008. *Journal of Cereal Science* 48:513–516.

Notes: DAA: days after anthesis
  Temperature regimes:
  1. 20/16°C (day/night) throughout (control)
  2. 20/16°C except for 35/20°C (2001) and 40/20°C (2002) for 72 hr at 25DAA
  3. 20/16°C with 16/25°C from 25DAA to maturity.
[a] LSD values for % UPP of the three treatments over 2 years were less than 0.5%.
[b] LSD value for SDS-sedimentation test was 0.12 mL.

the balance of dough properties toward more "bucky" doughs (high dough strength and low extensibility), as illustrated in Figure 11.12.

Heat stress

Crops subjected to high temperatures prior to harvest often respond with losses of quality as well as losses of yield (Finney and Fryer 1958; Blumenthal, Barlow, and Wrigley 1993). Table 11.8 summarizes some data for a pair of near-isogenic wheat lines grown under three temperature regimes over 2 years (Irmak et al. 2008). Plants were grown in the greenhouse and transferred to growth chambers for the heat treatments. UPP was used to monitor polymerization of glutenin, which begins to increase at a time intermediate between anthesis and maturity.

Three main conclusions resulted from the study. First, glutenin in the line with the *Glu-D1d* allele (HMW-GS 5+10) began to polymerize at an earlier time (by 3–6 days) than the line with the *Glu-D1a* allele (HMW-GS 2+12) and reached a higher value of UPP at maturity. Second, the application of heat stress caused a decrease in UPP and thus in MW for the two lines, suggesting that heat stress has a negative effect on glutenin polymerization. Third, the line with the strength-associated HMW-GS (5+10) appeared to be more tolerant to heat stress. Thus, even though this may not have been the primary aim of the study, information on the compositional changes was obtained by measuring the effect of varying environmental conditions while keeping the genetic composition constant.

Exercises

1. How does the magnitude of change in dough strength compare between the effects of (a) deletion of one HMW-GS and (b) allelic variation from HMW-GS 5+10 to HMW-GS 2+12 at *Glu-D1*? Use data from Lawrence et al. (1988) and Gupta and MacRitchie (1994) to attempt to answer.
2. What can be deduced about the ratio of LMW-GS/gliadins at the *Gli-B1/Glu-B3* locus of the parent lines from Table 11.4?

References

Ammar, K., A. J. Lukaszewski, and G. M. Banowetz. 1997. Effect of *Glu-D1* (5+10) on gluten strength and polymeric protein composition in durum wheat. *Cereal Foods World* 42:610.

Blechl, A. E., J. W. Lin, S. R. Nguyen, O. D. Anderson, and F. M. Dupont. 2007. Transgenic wheats with elevated levels of Dx5 and/or Dy10 high-molecular-weight glutenin subunits yield dough with increased mixing strength and tolerance. *Journal of Cereal Science* 45:172–183.

Blumenthal, C., E. W. R. Barlow, and C. W. Wrigley. 1993. Growth environment and wheat quality: The effect of heat stress on dough properties and gluten proteins. *Journal of Cereal Science* 19:3–21.

Finney, K. F., and M. A. Barmore. 1948. Loaf volume and protein content of hard red spring wheats. *Cereal Chemistry* 25:291–312.

Finney, K. F., and H. C. Fryer. 1958. Effect on loaf volume of high temperatures during the fruiting period of wheat. *Agronomy Journal* 50:28–34.

Gianibelli, M. C., O. R. Larroque, P. Chan, B. Margiotta, E. Deambrogio, F. MacRitchie, and D. Lafiandra. 1998. Effect of allelic variants at the *Gli-1/Glu-3* and *Gli-2* loci on mixing properties. In *Proceedings of the 9th International Wheat Genetics Symposium*, vol. 4, 154–156. ed. A. E. Slinkard. Saskatchewan, Canada: University of Saskatchewan.

Gupta, R. B., I. L. Batey, and F. MacRitchie. 1992. Relationships between protein composition and functional properties of wheat flours. *Cereal Chemistry* 69:125–131.

Gupta, R. B., and F. MacRitchie. 1994. Allelic variation at glutenin subunit and gliadin loci, *Glu-1*, *Glu-3*, and *Gli-1* of common wheats. II. Biochemical basis of the allelic effects on dough properties. *Journal of Cereal Science* 19:19–29.

Gupta, R. B., F. MacRitchie, K. W. Shepherd, and F. Ellison. 1991. Relative contributions of LMW and HMW glutenin subunits to dough strength and dough stickiness of bread wheat. In *Gluten proteins 1990*, ed. W. Bushuk and R. Tkachuk, 71–80. St. Paul, MN: American Association of Cereal Chemists.

Gupta, R. B., Y. Popineau, J. Lefebre, M. Cornec, and F. MacRitchie. 1995. Biochemical basis of flour properties in bread wheats. II. Changes in polymeric protein formation and dough/gluten properties associated with the loss of low M_r and high M_r glutenin subunits. *Journal of Cereal Science* 21:103–116.

Gupta, R. B., and K. W. Shepherd. 1993. Production of multiple wheat-rye 1RS translocation stocks and genetic analysis of LMW subunits of glutenin and gliadins using these stocks. *Theoretical and Applied Genetics* 85:719–728.

Irmak, S., H. A. Naeem, G. L. Lookhart, and F. MacRitchie. 2008. Effect of heat stress on wheat proteins during kernel development in wheat near-isogenic lines differing at *Glu-D1*. *Journal of Cereal Science* 48:513–516.

Lafiandra, D., R. D'Ovidio, E. Porceddu, B. Margiotta, and G. Colaprico. 1993. New data supporting high M_r glutenin subunit 5 as the determinant of quality differences among the pairs 5+10 vs. 2+12. *Journal of Cereal Science* 18:197–206.

Lafiandra, D., B. Margiotta, G. Colaprico, S. Masci, M. R. Roth, and F. MacRitchie. 2000. Introduction of the D genome related high- and low-M_r glutenin subunits into drum wheat and their effect on technological properties. In *Wheat gluten*, ed. P. R. Shewry and A. S. Tatham, 51–54. London: Royal Society of Chemistry.

Larroque, O. R., M. C. Gianibelli, I. L. Batey, and F. MacRitchie. 1997. Electrophoretic characterization of fractions collected from gluten protein extracts subjected to size-exclusion high-performance liquid chromatography. *Electrophoresis*. 18:1064–1067.

Lawrence, G. J., F. MacRitchie, and C. W. Wrigley. 1988. Dough and baking quality of wheat lines deficient for glutenin subunits controlled by *Glu-A1*, *Glu-B1*, and *Glu-D1* loci. *Journal of Cereal Science* 7:109–112.

Lawrence, G. J., H. J. Moss, K. W. Shepherd, and C. W. Wrigley. 1987. Dough quality of biotypes of 11 Australian wheat cultivars that differ in high-molecular-weight glutenin subunit composition. *Journal of Cereal Science* 6:99–101.

MacRitchie, F. 1978. Differences in baking quality between wheat flours. *International Journal of Food Science Technology* 13:187–194.

———. 1999. Wheat proteins: Characterization and role in flour functionality. *Cereal Foods World* 44:188–193.

MacRitchie, F., and P. W. Gras. 1973. The role of flour lipids in baking. *Cereal Chemistry* 50:292–302.

MacRitchie, F., and R. B. Gupta. 1993. Functionality–composition relationships of wheat flour as a result of variation in sulfur availability. *Australian Journal of Agricultural Research* 44:1767–1774.

Masci, S., D. Lafiandra, E. Porceddu, and D. D. Kasarda. 1993. D-glutenin subunits: N-terminal sequences and evidence for the presence of cysteine. *Cereal Chemistry* 70:581–585.

Paux, E., P. Sourdille, J. Salse, C. Saintenac, F. Choulet, P. Leroy, A. Korol, M. Michalak, S. Kianian, W. Spielmeyer, et al. 2008. A physical map of the 1-gigabase bread wheat chromosome 3B. *Science* 322:101–104.

Payne, P. I., K. G. Corfield, and J. A. Blackman. 1981. Correlation between the inheritance of certain high-molecular-weight subunits of glutenin and bread-making quality in progenies of six crosses of bread wheat. *Journal of the Science of Food and Agriculture* 32:51–60.

Payne, P. I., L. M. Holt, K. Harinder, D. P. McArtney, and G. J. Lawrence. 1987. The use of near-isogenic lines with different HMW glutenin subunits in studying bread-making quality and glutenin structure. In *Proceedings of the 3rd International Workshop on Gluten Proteins,* 216–226. ed. R. Lasztity and F. Bekes. Singapore: World Scientific Publishing.

Payne, P. I., L. M. Holt, and P. G. Lister. 1988. *Gli-A3* and *Gli-B3,* two newly designated loci coding for omega-type gliadins and D subunits of glutenin. In *Proceedings of the 7th International Wheat Genetics Symposium,* ed. T. E. Miller and R. M. D. Koebner, 999–1002. Cambridge: IPSR.

Pennisi, E. 2008. Edward Buckley profile: Romping through maize diversity. *Science* 322(5898):40–41.

Pogna, N. E., R. M. Redaelli, P. Vaccino, A. M. Cardi, A. D. B. Peruffo, A. Curioni, E. V. Metakovsky, and S. Pagliaricci. 1995. Production and genetic characterization of near-isogenic lines in the bread-wheat cultivar Alpe. *Theoretical and Applied Genetics* 90:650–658.

Rakszegi, M., G. Pastori, H. D. Jones, F. Bekes, B. Butow, L. Lang, Z. Bedo, and P. R. Shewry. 2008. Technological quality of field grown transgenic lines of commercial wheat cultivars expressing the 1Ax1 HMW glutenin subunit gene. *Journal of Cereal Science* 47:310–321.

Ruiz, M., and J. M. Carrillo. 1993. Linkage relationships between prolamin genes of chromosome 1A and 1B of durum wheat. *Theoretical and Applied Genetics* 87:353–360.

Sears, E. R. 1954. The aneuploids of common wheat. *Missouri Agriculture Experimental Station Research Bulletin* 572:1–58.

———. 1966. Nullisomic-tetrasomic combinations in hexaploid wheat. In *Chromosome manipulation and plant genetics,* ed. R. Riley and K. R. Lewis. Edinburgh: Oliver and Boyd.

Sreeramulu, G., and N. K. Singh. 1997. Genetic and biochemical characterization of novel low molecular weight glutenin subunits in wheat (*Triticum aestivum* L.). *Genome* 40:41–48.

Wrigley, C. W., D. L. Du Cros, J. G. Fullington, and D. D. Kasarda. 1984. Changes in polypeptide composition and grain quality due to sulfur deficiency in wheat. *Journal of Cereal Science* 2:15–4.

Yamamori, M., and N. T. Quynh. 2000. Differential effects of Wx-A1, -B1, and -D1 protein deficiencies on apparent amylose content and starch pasting properties in common wheat. *Theoretical and Applied Genetics* 100:32–38.

chapter twelve

Strategies for targeting specific end-use properties

Background

Cereal breeders usually focus their efforts on two main targets: high yield and high disease resistance. Functional (including nutritional) quality is a third target that needs input from cereal chemists in collaboration with breeders. Before considering how strategies may be applied in order to modify functionality in breeding programs, it is useful to trace some of the history of how cereal chemists have uncovered relationships between composition and end-use properties. Again, wheat will be used as the example. Although a few studies will be highlighted, it should be recognized that progress in any scientific endeavor results from the combined efforts of many.

In the 1940s, Finney pioneered fractionation and reconstitution techniques for wheat flours (Finney 1943) and showed that gluten protein was responsible for dough strength and bread-making performance. In the 1970s, Orth and Bushuk (1972) discovered that the amount of residue protein (insoluble in lactic acid) in flour gave a measure of bread-making potential. The 1970s decade was also a time when poor-quality wheat varieties appeared on the scene in Europe. These wheats had low milling quality and the weak properties of the dough from them gave processing problems in bakeries and poor bread-making performance. The problems arose because varieties were largely developed on the basis of yield potential. Booth and Melvin (1979) used fractionation and reconstitution techniques to study one of these varieties, Maris Huntsman. They found that its residue protein was much lower in amount than that of a typical, high-quality Canadian Western Red Spring (CWRS) wheat. Furthermore, interchange of protein fractions showed a lack of quality in lactic acid soluble and insoluble fractions.

Opportunity for input from cereal chemists

The processing problems associated with these high-yielding wheat varieties drew attention to the need to consider functional properties in breeding programs and provided an opportunity for cereal chemists to have

Table 12.1 Association of HMW-GS Alleles with Dough Strength

Locus	Strong dough	Weak dough
Glu-B1	7+8, 17+18	20x+20y
Glu-D1	5+10	2+12

an input. Work by Payne in the 1980s showed the important contribution of a protein component: the high molecular weight glutenin subunits (HMW-GS). As we have seen in Chapters 10 and 11, these subunits participate in building the large glutenin polymers responsible for the strength and elasticity of dough. Although they make up only about 10–15% of the total protein, their effect on functionality appears to be more than proportional to their amount.

In a series of publications, Payne and co-workers showed that the different HMW-GS present in wheat varieties as a result of allelic variation contribute to variations in functionality (Payne et al. 1981). Some HMW-GS, when present in varieties, were associated with high dough strength and good bread-making performance; others were associated with weak dough properties. Scores were assigned to individual HMW-GS and their combinations, based on studies of a large number of varieties in which the specific HMW-GS were identified by sodium dodecyl sulfate (SDS) polyacrylamide gel electrophoresis (PAGE) and related to the bread-making performance. These scores can be found in the literature (Payne 1987). Table 12.1 shows HMW-GS at the *Glu-B1* and *Glu-D1* loci that are well established as having contrasting contributions to dough strength.

The incorporation of strength-associated HMW-GS into varieties in European breeding programs had the effect of moving dough properties toward greater strength with consequent improvements in bread-making quality. In some cases, this led to overly strong dough having excessive mixing requirements (Wooding et al. 1999), thus creating processing problems in bakeries. Another effect of increasing dough strength too much (measured, for example, by R_{max}) is that dough extensibility may be impaired. The lesson is therefore that a systematic approach is needed to manipulate allelic composition in order to tailor functional properties to required end uses. Different properties need to be monitored to ensure that changing one parameter does not seriously affect another.

Relating allelic composition to end-use properties

Some traits associated with aspects of quality are dependent on only one or a few genes. For example, as we have seen in Chapter 4, grain hardness is controlled by a major gene on the short arm of chromosome 5D. The difference between red wheat and white wheat is caused by three genes that control bran pigments in red wheat. The percentage of amylose in wheat starch depends on three genes that code for the granule-bound starch synthase enzymes (GBSS) involved in amylose synthesis. Resistance to specific pathogens can often be related to one or a few genes.

In the case of physical properties, such as those important in processing (e.g., dough strength and extensibility, product quality), the situation is a bit different. These properties are dependent on the gluten proteins and how these proteins contribute to the two main variables that determine physical properties—namely, the ratio of polymeric proteins (mainly glutenins) to gliadins and the molecular weight distribution (MWD) of the polymeric proteins. Physical dough properties can be different as a result of variations in one or usually both of these parameters. Figure 12.1 illustrates

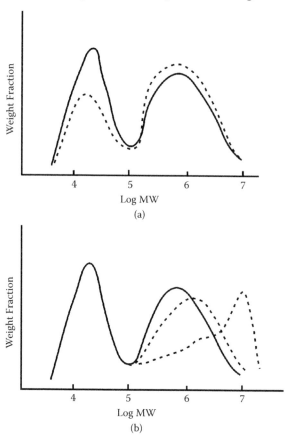

Figure 12.1 Schematic illustration of how the MWD of gluten protein can change by (a) a change in the ratio of glutenin/gliadin or (b) a change in the MWD of the glutenin.

how the two parameters can change. Because all of the proteins contribute to these parameters, processing quality is a polygenic characteristic.

The path from genetic composition to measures of processing quality can be summarized by the diagram of Figure 12.2. The proteins present in a wheat sample are determined by the different alleles that are present. The full protein complement then contributes to the two compositional variables (polymeric protein/gliadin ratio and MWD of the polymeric protein), which in turn define the functional properties. This statement may need to be qualified in the case of product quality. As we have seen from the discussion in Chapter 7, the gluten proteins only explain a part, albeit an important part, of bread-making performance. Another "joker in the pack," of course, is the effect of environmental variables. These can cause alterations of the final properties from those that were predicted on the basis of genetic composition.

Based on the scheme illustrated in Figure 12.2, it is possible, at least in theory, to design a wheat variety with specific properties by selecting the complete allelic composition controlling the proteins if we know the effects of all the alleles (proteins) on protein composition and thus the functional properties that we wish to achieve. These relationships were what we were pursuing in Chapter 11. Of course, the usual situation in a breeding program is not to have to start from scratch, but rather to take a current variety and modify its allelic composition in a way that will move the properties closer to those desired.

Let us suppose that we wish to modify the balance of dough properties in a variety. In cereal labs, load-deformation instruments such as the extensigraph and alveograph are commonly used to measure two main parameters that characterize dough properties: the resistance to extension (or tenacity) and the extensibility. Although measured by the same instruments, these two parameters appear to depend quite differently on protein composition. In a study based on a large number of wheat samples

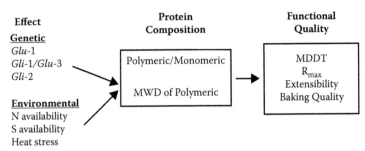

Figure 12.2 Schematic representation of how allelic composition (modified by environmental effects) determines the protein composition and consequently the functional properties.

(158), Bangur et al. (1997) found that extensigraph R_{max} gave the highest correlation with a fraction of the polymeric protein above a critical MW, estimated to be about 250,000. This fraction comprised about 60% of the total polymeric protein.

On the other hand, extensibility gave the highest correlation with the total percentage of flour polymeric protein. Other studies (Larroque, Gianibelli, and MacRitchie 1999) suggest that extensibility is impaired if the MWD of the glutenin is shifted to too high values, which is what would be expected intuitively. In order to increase dough strength, we therefore need to shift the MWD to higher values so that a greater percentage of polymeric protein will be above the critical MW and the MWD of this fraction will be enhanced. In terms of protein composition, the choices for achieving this are to introduce strength-associated high molecular weight glutenin subunit (HMW-GS) alleles, increase the HMW/low molecular weight (LMW)-GS ratio, and eliminate proteins behaving as chain terminators.

Of course, these changes will tend to have a negative effect on dough extensibility. Therefore, care must always be taken to ensure that, by improving one quality parameter, another one is not penalized. A large variation in the ratio of R_{max} to extensibility is possible by manipulating the two main compositional variables of Figure 12.2. For example, if we require dough properties with a very high extensibility but a moderate R_{max}, we can increase the percentage of polymeric protein in the flour while preventing the MWD of this protein from shifting to values that are too high. The ratio of polymeric proteins to gliadins is, to an approximation, determined by the number of glutenin subunits relative to the number of gliadins. Thus, alleles that have a large number of genes coding for glutenin subunits and a small number of genes coding for gliadins would need to be selected (cf. Singh, Donovan, and MacRitchie 1990).

To illustrate how we might apply the strategy outlined previously, we will look at one practical problem in the area of wheat processing quality.

Wheat/rye translocation lines

Wheat/rye translocation lines were briefly discussed in Chapter 11. The most common ones and the ones present in some commercial wheat consignments are the 1BL/1RS lines. These are lines in which the short arm of chromosome 1B of wheat has been substituted by the short arm of chromosome 1R of rye. This is shown in Figure 12.3. The effect of the translocation is to eliminate the *Gli-B1/Glu-B3* locus. Portions of the gamma- and omega-gliadins as well as the LMW-GS are thus deleted.

The gliadins are, to some extent, replaced because the short arm of chromosome 1 of rye introduces secalins, coded at the *Sec-1* locus. These

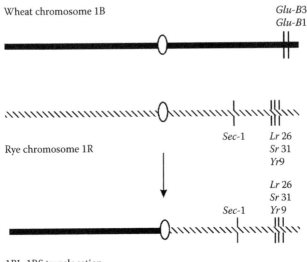

Wheat chromosome 1B

Glu-B3
Glu-B1

Sec-1   Lr 26
         Sr 31
         Yr9

Rye chromosome 1R

         Lr 26
         Sr 31
Sec-1   Yr 9

1BL-1RS translocation

Figure 12.3 1B/1R wheat/rye translocation.

secalins, like gliadins, are single chain or monomeric proteins. Another important change is that genes for resistance to a wide range of pathogens (stem rust, leaf rust, and stripe rust) are introduced. This makes these lines attractive agronomically. Unfortunately, their functional quality is usually very poor. Their doughs tend to be weak and sticky, giving processing problems in bakeries and performing poorly in bread-making. It would therefore be valuable to be able to eliminate the dough stickiness problem in order to utilize the enhanced disease resistance that they offer. This may serve as a model for problems in functionality caused by introduction of alien genes.

Let us see how we might apply the approach shown in Figure 12.2 to seek a solution to the problem. In terms of functionality, the greatest effect of the translocation is to eliminate the *Glu-B3* locus, coding for LMW-GS, which are more abundant than HMW-GS in a normal wheat (usually by a factor close to three). If we consider that protein from normal wheat has about 40% glutenin, then the loss of the *Glu-B3* locus will reduce the glutenin content to about 30%, assuming that each of the *Glu-3* loci contributes equally to the LMW-GS.

An interesting result is shown in Table 12.2. Values of R_{max} are compared for flours from a soft wheat parent variety, Grebe, and its 1B/1R translocation line. Surprisingly, R_{max} is slightly greater for the translocation line, although the 1B/1R line has the greater stickiness. It has been suggested that one solution to the unsuitable dough properties of 1B/1R lines might be to utilize parent lines with alleles for extra strong

Table 12.2 Dough Strength and Stickiness Data for a Soft Wheat and Its 1B/1R Translocation

Variety	R_{max} (BU)	Dough stickiness
Grebe parent	220	Not sticky
Grebe 1B/1R translocation	260	Sticky

HMW-GS. However, the results of Table 12.2 (F. MacRitchie, unpublished results) suggest that lack of gluten strength may not be the major problem. The effect of eliminating the *Glu-B3* locus is to shift the HMW/LMW-GS to a higher value. This may explain the higher value of R_{max} for the 1B/1R line above that of the relatively weak soft wheat parent Grebe. An increase of the HMW/LMW-GS ratio, as we have seen (Chapter 11), would be expected to shift the MWD to higher MW and therefore lead to greater dough strength.

The factor that may be mainly responsible for inducing sticky dough properties is the reduction in the glutenin/gliadin ratio or, more generally, the polymeric/monomeric protein ratio because secalins are also involved (Dhaliwal and MacRitchie 1990). This is supported by experiments in which this ratio is deliberately varied by mixing fractions concentrated in either glutenin or gliadin. As the proportion of gliadin is increased, sticky dough properties are encountered. A normal bread wheat flour has a glutenin/gliadin ratio close to unity. Elimination of the *Glu-B3* locus reduces this ratio to the order of 30/40 or 0.75, a considerable change.

A solution to the 1B/1R processing problem would therefore require a means of maintaining a high glutenin/gliadin ratio without altering markedly the HMW/LMW-GS ratio. The parent line would thus need to be one in which this ratio (glutenin/gliadin) is unusually high (say, 1/0.75 or 1.33) so that loss of one *Glu-3* locus would only lower the glutenin/gliadin ratio to close to unity. Thus, a parent line with a high number of genes coding for glutenin subunits and a relatively low number of genes coding for gliadins would be required. In the absence of suitable parents, approaches that might be used to obtain such lines include the use of *Gli-2* null lines (see Chapter 11) or mutagenesis.

Methods for manipulating genes

It is useful here to mention briefly some of the technology that is being developed to manipulate genes in order to modify functionality. Gene transfer technology (transformation) was mentioned in Chapter 11 as a means for determining protein composition–functionality relationships. In principle, it can also be used in plant breeding. Until now, its application has been most promising for conferring disease resistance to some cereals, notably rice and maize.

Gene silencing is an exciting area of study because it has the potential to "switch off" specific genes in diseased tissue. In plants, post-transcriptional gene silencing involves the use of RNA interference (RNAi), a mechanism that uses small interfering RNAs (siRNAs) to target specific RNAs so that they can no longer be translated into protein. An earlier study (Waterhouse, Graham, and Wong 1998) and a recent one applied to rice (Warthmann et al. 2008) provide useful background for those not conversant with the topic.

Mutagenesis was mentioned in the previous section. Targeting induced local lesions in genomics (TILLING) is a promising method for inducing and identifying novel genetic variation (Gilchrist and Haughn 2005). Initially, it has been used for research on *Arabidopsis;* however, with the progress made in genomics (briefly mentioned in Chapter 11), its application to many crops, including cereals, is becoming more feasible. Novel single base-pair changes are induced in a population of plants by treating seeds (or pollen) with a chemical mutagen. The chemicals that have been used most are ethylmethane sulfonate (EMS) and ethylnitrosourea (ENU). The use of a mismatch specific celery nuclease (CEL1) to identify single nucleotide polymorphisms (SNPs) from the polymerase chain reaction (PCR)-amplified DNA enables a high throughput of samples to be achieved. The procedure and its potential application are well described by Slade and Knauf (2005).

Exercise

1. Dough stickiness is a problem that can also occur in wheat lines that do not have rye translocations. An Australian variety, Halberd, is well known for having been associated with sticky dough, although it had good agronomic characteristics and relatively high dough strength, measured by R_{max}. Some information about the protein composition of Halberd is given by Singh et al. (1990) and Gupta, Batey, and MacRitchie (1992). Use this information to attempt an explanation of the sticky dough properties of this variety.

References

Bangur, R., I. L. Batey, E. McKenzie, and F. MacRitchie. 1997. Dependence of extensigraph parameters on wheat protein composition measured by SE-HPLC. *Journal of Cereal Science* 25:237–241.

Booth, M. R., and M. A. Melvin. 1979. Factors responsible for the poor bread-making quality of high yielding European wheat. *Journal of the Science of Food and Agriculture* 30:1057–1064.

Dhaliwal, A. S., and F. MacRitchie. 1990. Contribution of protein fractions to dough handling properties of wheat/rye translocation cultivars. *Journal of Cereal Science* 12:113–122.

Finney, K. F. 1943. Fractionating and reconstituting techniques as tools in wheat flour research. *Cereal Chemistry* 20:381–396.

Gilchrist, E. J., and G. W. Haughn. 2005. TILLING without a plough: A new method with applications for reverse genetics. *Current Opinion in Plant Biology* 8:211–215.

Gupta, R. B., I. L. Batey, and F. MacRitchie. 1992. Relationships between protein composition and functional properties of wheat flours. *Cereal Chemistry* 69:125–131.

Larroque, O., M. C. Gianibelli, and F. MacRitchie. 1999. Protein composition for pairs of wheat lines with contrasting dough extensibility. *Journal of Cereal Science* 29:27–31.

Orth, R. A., and W. Bushuk. 1972. A comparative study of the proteins of wheats of diverse baking qualities. *Cereal Chemistry* 49:268–275.

Payne, P. I. 1987. The genetical basis of bread-making quality in wheat. *Aspects of Applied Biology* 15:79–90.

Payne, P. I., K. G. Corfield, L. M. Holt, and J. A. Blackman. 1981. Correlations between the inheritance of certain high-molecular weight subunits of glutenin and bread-making quality in progenies of six crosses of bread wheat. *Journal of the Science of Food and Agriculture* 32:51–60.

Singh, N. K., R. Donovan, and F. MacRitchie. 1990. Use of sonication and size-exclusion high-performance liquid chromatography in the study of wheat flour proteins. II. Relative quantity of glutenin as a measure of bread-making quality. *Cereal Chemistry* 67:161–170.

Slade, A. J., and V. C. Knauf. 2005. TILLING moves beyond functional genomics into crop improvement. *Transgenic Research* 14:109–115.

Warthmann, N., H. Chen, S. Ossowoski, D. Weigel, and P. Herve. 2008. Highly specific gene silencing by artificial mRNAs in rice. *PLoS* ONE Mar 19; 3(3):e1829.

Waterhouse, P. M., M. W. Graham, and M. B. Wong. 1998. Virus resistance and gene silencing in plants can be induced by simultaneous expression of sense and antisense RNA. *Proceedings of the National Academy of Sciences U.S.A.* 95:13959–13964.

Wooding, A. R., S. Kavale, F. MacRitchie, and F. L. Stoddard. 1999. Link between mixing requirements and dough strength. *Cereal Chemistry* 76:800–806.

chapter thirteen

Nonwheat cereals

Introduction

As stated in Chapter 1, wheat is the cereal that has attracted by far the most research efforts. A good deal of this research has adopted theory from the basic sciences and more conceptual ideas have emerged for wheat than for the other cereals. Nevertheless, world production and utilization of rice and maize (corn) are of the same order as that for wheat.

Other cereals fill specific needs. For example, rye is the only cereal that approaches wheat in its capacity for bread-making (although it is much inferior), and rye bread has high acceptance in certain locations, especially northern Europe. Barley has been the cereal of choice for malting and brewing. Sorghum (also millets) is a cereal that can be cultivated in areas that are unsuitable for growing other cereals, especially areas that experience drought stress. It is grown extensively in Africa, where it is utilized for food for humans. In recent times, there has been greater recognition of certain intolerances to gluten in the diet of a section of the population (see Chapter 14).

Also, the presence of compounds beneficial to health (phytochemicals) has been better elucidated, and these are found in greater concentrations in cereals such as barley and oats. As a result of these new findings, research is tending to shift toward greater efforts to study nonwheat cereals. This is reflected in an increasing proportion of publications on these cereals. Some of the areas where conceptual ideas are being applied or may be needed will be briefly mentioned in this chapter.

Rice

Distribution of nutrients in kernel

Milling of rice is different from that of some of the other cereals in that there is usually no grinding step to produce a powder. Milling involves removal of the hull and bran layers of the rough rice kernel (paddy). Hulled rice (brown rice) is composed of surface bran (6–7% by weight), endosperm (about 90%), and embryo (2–3%). Milling of brown rice uses abrasion and friction between kernels to produce polished or whitened rice, where some 8–10% of the mass (mainly bran) has been removed.

Increasing the degree (intensity and time) of milling results in removal of increasing amounts of bran.

Like most cereals, rice kernel structure and composition vary from its outer to inner sections. Because nutrients are more concentrated in the outer layers of the kernel, milling can cause a substantial loss of nutrients as well as a change in the edible properties of the rice. Consequently, information on the distribution of nutrients is being sought to help understanding of how the milling process can be optimized to improve sensory properties while retaining essential nutrients as much as possible. There are worldwide deficiencies of certain minerals and their bioavailability may be impaired by formation of insoluble mineral–phytate complexes, as discussed in Chapter 14.

Liang et al. (2008) have analyzed zinc by atomic absorption and phytic acid by spectrochemical detection of phosphorus. They used x-ray fluorescent microscopy imaging to visualize the distribution of zinc and phytic acid in kernels of several rice varieties by making use of increasing degrees of abrasive milling. It was concluded that milling characteristics, including mass loss and kernel breakage, varied among rice cultivars that had differently shaped kernels.

For the cultivars studied, phytic acid was found to be concentrated in the outermost layer. Zinc distribution, on the other hand, was similar for three cultivars and was characterized by an even distribution throughout the kernel, with the exception of a higher concentration in the embryo. Studies of this nature will be valuable for optimizing the milling process for maximum removal of phytic acid with minimum mineral and yield loss and appropriate whiteness to satisfy consumer expectations for white rice.

Shelf life of cooked rice

Demand for ready-to-eat meals is increasing and consumption of rice after cooking and storage is becoming more popular, especially in Asia. The "keeping" quality is therefore an important parameter. Many studies have reported different factors influencing quality of stored cooked rice. Some of the factors studied have been the effects of cultivar (Singh et al. 2005), starch structure (Ong and Blanshard 1995), and storage temperature (Perdon et al. 1999). The introduction of cereal varieties with a wide range of amylose/amylopectin (see Chapter 11) is proving valuable for studying the effects of starch composition on different functional properties.

A recent study by Yu et al. (2009) used rice varieties with amylose contents ranging from 1.2 to 35.6% to follow the textural properties of hardness and adhesiveness of cooked rice during storage at low temperature (4°C). Rice with high amylose content retrograded rapidly during storage, whereas rice with low amylose retrograded slowly. As a result, both hardness (measured by a texture analyzer) and the enthalpy of retrogradation, ΔH_r (measured by DSC), correlated highly positively with the proportion of amylose. Adhesiveness, measured by repeat cycles of compression, correlated negatively with amylose. It was suggested that amylose retrograded within 1 day, producing crystal nuclei, which increase the rate of starch crystal growth. Thus, for rice with high amylose, the starch retrograded faster, giving higher ΔH_r values.

Grain chalkiness

Chalkiness in rice grains occurs as a result of loose packing of amyloplasts, the organelles in cells that are responsible for the synthesis and storage of starch granules. Chalky grains have a negative impact on quality. When the proportion of chalky grains exceeds about 15%, rice has a decreased eating quality (Kim et al. 2000). As a result, the value of rice is decreased in the world market. High-temperature stress (above 26 or 27°C) during grain ripening increases the formation of chalky grains. The problem has been observed to increase in recent times and, in view of the predictions for global warming, it is important that it be addressed.

The physiological origin of chalky grains has not been well elucidated. Several hypotheses have been advanced:

insufficient supply of nutrients to the developing endosperm (Sato and Inaba 1976)
reduced ability to synthesize starch in the endosperm (Jiang, Diane, and Wu 2003; Lin et al. 2005)
degradation of starch by α-amylase during ripening (Yamakawa et al. 2007)

Several complementary techniques have been used by Ishimaru et al. (2009) to throw light on the phenomenon. The disorganized development of amyloplasts was observed with the use of scanning electron microscopy. The distribution of water was followed by magnetic resonance imaging. This technique, based on nuclear magnetic resonance, has been used for measuring the water distribution in developing grains of barley (Glidewell 2006) and rice (Horigane et al. 2001).

The study showed that, after the middle stage of grain ripening, the water content in the central chalky part became higher than in the lateral translucent part. It was suggested that loose packing of amyloplasts in the chalky part may allow pooling of water in the free space. Another technique that was applied was the use of transcripts for α-amylase. No m-RNAs for α-amylase were detected in the early and middle stages of grain ripening. This was taken as evidence contrary to the validity of the third hypothesis (degradation during ripening).

New approaches in breeding

Gene mapping has come to the forefront as a result of the development of rapid methods for analysis of DNA. It is now being applied for improvement of cereals in breeding programs. Genetic markers are genes or DNA sequences that can be traced to specific locations on chromosomes and associated with particular traits. A genetic marker must exhibit polymorphism; that is, it must have two or more common variations in the population studied.

Linkage maps are being constructed that give the order and genetic distances between genes for variable, single traits that are linked. This approach enables identification of quantitative trait loci (QTL), which are stretches of DNA closely linked to the genes that underlie the particular trait under study. The idea behind QTL mapping is to determine the degree of association of a specific region of the genome to the inheritance of the trait of interest.

An example of the application of QTL mapping of grain quality traits in rice has been described by Lou et al. (2009). By crossing rice varieties of small and large grain sizes, a set of recombinant inbred lines (RILs) was produced and used to identify QTLs controlling eight grain quality traits. The main components of rice grain quality that influence commercial values are appearance quality, milling quality, cooking–eating quality, and nutritional quality—each of which can, in turn, be subdivided into several characteristics.

Another interesting study of QTL mapping for rice characteristics using a doubled haploid population has been reported by Bao et al. (2004). A doubled haploid is a genotype formed when haploid plants derive two sets of identical chromosomes. The normal haploid cells undergo chromosome doubling as a result of induced or spontaneous processes, thus producing completely homozygous plants. These genotypes are precisely repeatable and hence heritability of quantitative characteristics is increased, leading to improved selection criteria.

Because the QTLs are small and highly influenced by environmental factors, accurate phenotyping with replicate trials is needed. This is possible by doubled haploidy because of their true breeding nature. Thus, whereas conventional breeding procedures take at least six generations to achieve a completely homozygous condition, doubled haploidy reduces this to one step. Bao et al. (2004) have mapped QTLs in the doubled haploid population of rice to study gelatinization temperature, gel textural traits, and flour swelling volume.

Maize (corn)

Dough viscoelasticity

One of the questions posed in Chapter 1 was "Why does dough from cereals other than wheat not have viscoelastic properties?" It is well established that the gluten proteins of wheat are responsible for the viscoelastic properties of wheat flour dough (see Chapter 6). The requirements for a protein (or any polymer) to exhibit viscoelasticity are discussed in Chapter 14. One of these is that the protein should be above its glass transition temperature (T_g). Zein, the prolamin protein of maize, is found to contribute viscoelastic properties to a zein–starch dough when the temperature is raised above its T_g at that water content (Bushuk and MacRitchie 1989; Lawton 1992). This shows that it is possible to obtain viscoelastic properties with nonwheat cereal proteins.

We therefore need to focus on how nonwheat cereal proteins differ from wheat proteins to seek an understanding of the question. The characterization of wheat proteins has progressed more than that of the nonwheat cereal proteins in recent times. The Osborne procedure is still used for characterizing proteins from nonwheat cereals; however, as we have seen in Chapter 9, although this method has been valuable, it does not give a reliable quantification of the different protein classes. In the case of wheat proteins, the application of sonication to solubilize the largest glutelins and size-exclusion high-performance liquid chromatography (SE-HPLC) to fractionate the protein classes gives more precise measurements.

Duplication of this procedure may not work so well for protein from other cereals, but it may serve as a basis for developing sounder methods. Milling of wheat also has the advantage of producing relatively pure flour free from bran and germ components that can negatively affect dough properties. An essential requisite for characterization is that the total protein content must be evaluated. Some of the attempts to measure protein composition of different cereals have neglected this simple requirement and can make the results unreliable. Lazstity (1996) has provided a good summary of the percentages of the main protein classes in maize that have been reported by independent researchers using the sequential solubilization procedure. These appear to show quite similar composition to wheat proteins—that is, roughly equal amounts of prolamins and glutelins (approximately 40% each) and similar contents of albumins/globulins (approximately 15%).

If these results can be relied on, it suggests that the different behavior of wheat and maize proteins may depend on differences in amino acid composition/structure of the functional proteins rather than differences in their relative proportions. How these differences might influence the T_g is an area that offers a challenge for future research. Of course, when seeking explanations for lack of dough viscoelasticity in nonwheat cereals, obvious variables need to be examined (e.g., relative proportions of prolamins and glutelins, MWD of the proteins, and presence of fiber compounds).

Quality protein maize

Cereals have low contents of essential amino acids, particularly lysine and tryptophan. In developing countries especially, this creates nutritional deficiencies for a large proportion of the population because a single cereal may account for a major part of the protein intake. In some countries (Central and South America and sub-Saharan Africa), people rely on maize as their principal food. In order to counter the problem, high-lysine mutants of maize have been identified. However, these mutants are associated with negative effects on yield and they have soft, chalky kernels that are susceptible to disease and insect damage.

The challenge of developing maize with high levels of lysine combined with good yield and hard kernels was taken up successfully by two researchers at CIMMYT (Centro Internacional de Mejoramiento de Maíz y Trigo [the International Maize and Wheat Improvement Center]), Mexico. The story of the collaboration between Villega, in charge of the laboratory investigating protein quality, and Vasal, a plant breeder, is a fine example of how patient, focused, multidisciplinary research can overcome a difficult problem. Vasal and Vallega combined to develop a mutant maize, opaque2, to produce quality protein maize (QPM) with hard-kernel characteristics and good taste similar to traditional maize, but with high levels of lysine and tryptophan.

An account of how the research evolved can be found in Prasanna et al. (2001). An updated review of the topic has been given by Gupta et al. (2009). The QPM maize germplasm has now been dispatched to different countries around the world and has had a great influence in improving the nutrition of a large number of people as well as animals.

Apart from the development of opaque2, other attempts to use high-lysine maize mutants using conventional breeding have not been very successful. Although lysine is increased, improvement of other grain characteristics, including yield, has not had much success. This has resulted in attempts to improve nutritional quality—not only in maize, but also in other cereals—by genetic manipulation. High-lysine maize mutants are characterized by decreases in synthesis of the prolamin proteins (zeins), which have very low lysine contents. Genetic engineering techniques are being applied to try to decrease zein synthesis without the accompanying deleterious effects on grain quality. The approaches that are being used have been well summarized by Shewry (2007).

Sorghum

Sorghum (and millets) are the most drought-tolerant cereal grain crops and require little input during their growth (Taylor, Schober, and Bean 2006). With increasing world population and decreasing water supplies, they therefore represent potentially important crops for future human use. However, although they are vital crops for millions of people in parts of Asia and Africa, they are relatively underutilized in the more developed countries. For example, the United States is a large producer of sorghum, but practically all of it is used for stock feed. Nevertheless, increasing research is in progress to utilize it for human food as well as for the production of ethanol and bioindustrial products such as bioplastics. Some of this research and the obstacles encountered have been discussed by Taylor et al. (2006).

Aerated products

Production of aerated products from nonwheat cereals has been a goal for a long time. Utilization of nonwheat cereals such as sorghum would be enhanced if they could produce breads with consumer acceptance close to that of wheat breads. This would also create a niche market for the fraction of the population that is gluten intolerant (see Chapter 14). Success has been limited.

Products that have been developed usually do not have high acceptability because of lower specific volumes, undesirable flavor, a high glycemic index if a large proportion of starch is used, and rapid staling. There is a need for a systematic fundamental approach to the quest for aerated products from the nonwheat cereals, especially those not causing health problems such as gluten intolerance. Some of the guidelines have been elucidated based on studies of aerated products from wheat; these are summarized in Chapter 14, where the requirements for viscoelastic doughs are discussed.

Sorghum proteins

The glutenins of wheat have similar amino acid composition to the gliadins. In some classifications, glutenins are considered to be glutelins, whereas in others (Shewry et al. 1986), they are classed as prolamins based on solubility of the gene products (glutenin subunits) and their similar genetic origin to gliadins. The essential difference between them is that gliadins are single chained (monomeric) proteins, whereas glutenins are multichained (polymeric) proteins in which the subunits are bonded together by disulfide bonds.

Classification of the nonwheat proteins has not been so clear. The prolamin proteins of sorghum are kafirins, which have a high degree of

sequence homology to maize proteins. They have been reported to consti-
tute 70–80% of the total protein in whole-grain sorghum flour (Hamaker et
al. 1995). Four classes of kafirin proteins have been identified: (1) α-kafirins,
(2) β-kafirins, and (3) γ-kafirins, based on solubility, electrophoretic mobility, and amino acid composition and sequences; and (4) δ-kafirins, based
on sequence of cloned DNA (Izquierdo and Godwin 2005). The sequential
solubility classification of sorghum proteins divides the proteins into

albumins
globulins
kafirins (prolamins, soluble in aqueous alcohol)
cross-linked kafirins (soluble in aqueous alcohol plus reducing agent)
cross-linked glutelins (soluble in detergent plus reducing agent plus
     alkaline pH)
unextracted structural protein residue

A newer and more simplified classification scheme has been proposed that divides them into two groups: kafirins and non-kafirins; the
non-kafirins comprise albumins, globulins, and glutelins (Hamaker
and Bugusu 2003). It can be concluded that classification of sorghum
proteins presents problems due to their difficulty in solubilizing.
Distinction between cross-linked kafirins and cross-linked glutelins
appears unsatisfactory, based simply on degree of solubility. There is
a need to concentrate research on developing a clearer classification of
sorghum proteins.

Nutritional aspects

Although sorghum is a principal source of energy and protein for millions, its nutritional value is diminished because of low digestibility of
grain protein and starch, which is exacerbated by cooking. Indigestibility
has been related to the amount of disulfide bonded (referred to as cross-
linked) proteins. In the sorghum endosperm, it is postulated that non-
kafirins form around protein bodies, effectively gluing the bodies into a
matrix surrounding the starch granules. This protein matrix is believed
to act as a barrier to starch gelatinization and digestibility due to cross-
linking between α- and β-kafirins and matrix proteins.

Wong et al. (2009) studied two sorghum lines with a common pedi-
gree that differed in digestibility. Indigestibility was shown to be related
to a greater abundance of disulfide bonded proteins. In the case of wheat,
much research has resulted in better characterization of the disulfide
bonded proteins (glutenins). There is a need to increase efforts to characterize the disulfide bonded proteins of maize and sorghum to lay the
foundations for breeding lines of enhanced end-use properties.

Millets

Millets comprise several species that, together with sorghum, are staple foods that supply a large proportion of calories and protein in semiarid tropical regions of Africa and Asia. They are roughly divided into two groups—pearl millet and small millet—based on seed size. Pearl millet is the most widely cultivated species. It is more efficient than sorghum or maize for utilization of moisture. Work by Zegada-Lizarazu and Iijima (2005) has shown that the drought resistance of pearl millet is explained by higher water-use efficiency rather than increased water uptake. This property of high water-use efficiency is likely to become a targeted trait for cereals in view of predicted climate changes and increase in drought stress in the future.

Less research has been carried out on millets than on other cereals. Nutritional quality is one area of priority. Like other cereals, millet has low contents of essential amino acids. For example, pearl millet contains in the order of 3 g lysine/100 g protein, whereas the World Health Organization recommends a minimum of 5.5 g/100 g. On the plus side, millets do not contain gluten proteins, so they are safe for those with celiac disease. Millets are low in certain minerals, including calcium.

Increasing the amounts of minerals and vitamins in the grain may not, however, be sufficient to improve the diet of consumers because they are concentrated in the pericarp, aleurone, and germ and are hence removed by decortification. It is therefore necessary to increase the concentration of vitamins and minerals in the endosperm and, in the case of minerals, their availability.

Genetic engineering techniques have had some success in improving quality for other cereals but their application to sorghum and millets is in the very early stages. Some of the main efforts to improve nutritional quality and resistance to pathogens and pests of sorghum and millets using biotechnology are being carried out at ICRISAT (International Crops Research Institute for the Semiarid Tropics).

Barley

Barley is one of the most ancient cereals. In early times, it was used as food for humans. Its utilization evolved into feed for stock and malting and brewing due in part to the growing importance of wheat and rice. In more recent times, there has been greater interest in its food potential for humans as a result of the health benefits of compounds present at higher levels than in other sources (see Chapter 14). These include fiber compounds such as β-glucans, which are known to lower cholesterol levels.

Barley is one of the most genetically diverse cereal grains. Genetic diversity provides opportunities to identify traits and breed varieties suitable for specific end uses. Some requirements for malting and brewing quality of barley have been established, but guidelines for processing quality of functional foods are lacking (Baik and Ulrich 2008). "Functional foods" is the term used for any food—fresh or processed—that is claimed to have a health-promoting or disease-preventing property beyond the basic function of supplying nutrients. Pursuit of knowledge about the factors that determine functionality in complex food systems is likely to be an active area of research in the future.

Processing quality

A large amount of variation in kernel hardness is found for barley varieties. Soft-grained varieties are preferred for malting and brewing, whereas hard-grained varieties tend to be favored for nonmalting applications. In hard-grained varieties, the fracture plane traverses the starch granule–protein interface similarly to hard-grained wheats (Brennan et al. 1996). This suggests a high degree of starch–protein adhesion. In the case of wheat, this difference has been ascribed to higher amounts of nonstick proteins at the surface of the starch granules of soft-grained wheat (see Chapter 4).

This has not been found to be the case for barley. Darlington et al. (2000) used a dry sieving procedure to isolate starch granules from both wheat and barley in order to avoid redistribution during water washing of starches. Higher levels of friabilin (the nonstick protein) were found to be associated with the starch granules of soft rather than hard wheat, but similar levels were found in the two types of barley. Hordoindolines, which are the homologues of puroindolines of wheat (Chapter 4), have been identified by Gautier et al. (2000). However, to date, no relationship between hordoindolines and grain hardness has been found. This apparent paradox needs to be resolved.

Another quality attribute is color. Barley grains are normally light yellow, but color can vary through purple, violet, blue, and black; these different colors are mainly caused by the presence of anthocyanins from the hull, pericarp, and aleurone layers (Edney et al. 1998). These colored types are attracting attention for application in functional foods due to the antioxidant properties of anthocyanins. Hulless barley contains 11–20% total dietary fiber, of which 11–14% is insoluble and 3–10% is soluble (Vikki et al. 2004). Pearling (removal of the outer layers by abrasion) reduces the contents of insoluble fiber, protein, ash, and lipids in the grain because these are more concentrated in the hull, bran, and germ. As a result, the pearled grain is richer in starch and β-glucans.

Beta-glucan enrichment

Barley contains relatively high amounts of soluble fiber (β-glucans). Beta-glucans are a diverse group of polysaccharides occurring as cell wall components in nature (Bacic, Stone, and Fincher 2009). They vary in molecular weight, solubility, and viscosity and, as discussed in Chapter 14, have well-established health benefits when included in the diet.

A method for producing barley flours enriched in β-glucans using micronization (reduction of particle size) and air classification has been reported by Ferrari et al. (2009). Fractions are prepared that enable a cumulative curve of β-glucan versus yield to be calculated. This allows the most suitable combination of yield and β-glucan content to be chosen. A study of two hulless barleys with different starch types produced barley flour fractions with twice the natural β-glucan concentration. Enriched fractions with 11.2 and 15.6% β-glucans were obtained with a flour yield of 30%.

Another approach to obtaining barley lines with high β-glucan content is through breeding. Li et al. (2008) have identified QTLs for β-glucan concentration in barley grain. A QTL explained up to 39% of the β-glucan concentration. Thus, genetic markers associated with the locus may be used to aid selection of high and low β-glucan barley lines.

Oats

Oats are classified broadly as hulled or naked (hulless); the naked oats are nutritionally superior. Naked oats have a thin, nonlignified husk on the outside of the grain. This falls off during harvesting, resulting in grain of higher energy, protein, and lipid, and lower fiber content compared to conventional oats. Oats have been an important human food crop in developing countries. In more developed economies, interest is increasing because of the dietary benefits associated with phytochemicals such as β-glucan. Together with barley, oats are the cereal with the highest amounts (3.0–5.0%) of β-glucans (Chernyshova et al. 2007).

Oats have a thicker layer of cell walls in the subaleurone region of the kernel than other cereals do. In addition to β-glucan, oats are a source of many beneficial health compounds (Peterson 2001), including antioxidants such as vitamin E (tocols), phytic acid, phenolic compounds, and avenanthramide; flavonoids and sterols are also present. Antioxidants assist in maintaining the stability of processed oat products. Oats can also stabilize oils and fats against rancidity.

It is likely that one direction of research in the future will be aimed at increasing the concentrations of health-related compounds. One approach is through breeding, where the breeder deals with genotype, environment, and cultivation procedures. Another is through cereal chemistry to devise methods of fractionation to concentrate the high-value compounds. Still another avenue is by milling and sieving (Gray et al. 2000). This will create opportunities for increased incorporation into breakfast foods, bread, beverages, and infant foods (Flander et al. 2007). In the development of oat products, processors need to be concerned not only with the level of compounds such as β-glucan, but also with properties like molecular weight distribution and solubility and how these are affected by variables such as temperature and enzyme activity.

Rye

Rye is mainly produced and consumed as bread in northern Europe. It is the only nonwheat cereal that is used extensively in bread-making and, similarly to wheat, suffers from the problem of gluten intolerance for a section of the population. However, because rye bread is normally made from whole meal, especially in Nordic countries, it contains a high dietary fiber content as well as myriad compounds that have been associated with health benefits. Although epidemiological studies provide evidence that whole grains are protective against diseases, the role of individual whole grain components is not well established.

Like other cereals, such as barley and oats, a good deal of the research on rye is being advocated for identifying the health-related compounds and trying to understand their role in metabolic processes (Bondia-Pons et al. 2009). Rye contains a high content of phenolic compounds that comprise a number of subgroups, including phenolic acids, flavonoids, isoflavonoids, lignans, stillbenes, and complex phenolic polymers (Heinio et al. 2008). Most are concentrated in the outermost aleurone and bran layers.

Although rye is used for bread-making, its dough viscoelastic properties are much inferior to those of wheat. Nevertheless, because its proteins do show properties more similar to wheat than do other nonwheat cereals, a detailed study of them seems to be justified to determine how they differ from wheat proteins. This could lead to better understanding of the requirements for dough viscoelasticity of nonwheat cereals.

Triticale

Triticale is a hybrid of wheat (*Triticum*) and rye (*Secale*) that was first bred in labs during the late nineteenth century. When wheat and rye are crossed, wheat is used as the female parent and rye as the male parent (pollen donor). The resulting hybrid is sterile and has to be treated with the alkaloid chemical colchicine to make it fertile so that it can reproduce. Different ploidy levels have been created (i.e., different numbers of chromosomes), but hexaploid triticale has been the most successful.

The original aim in developing triticale was to combine the disease resistance and environmental tolerance (soil, water, stress) of rye with the yield potential and bread-making properties of wheat. So far, this aim has only been partly realized. A program was initiated at CIMMYT in 1964 to improve triticale quality. This has resulted in quite spectacular increases in yield. For example, at Obregon, Mexico, in 1968, the highest yielding triticale produced 2.4 t/ha. Today, lines have exceeded 10 t/ha under optimum conditions.

Conventional breeding has established triticale as a valuable commercial crop, mainly for stock feed, particularly where conditions are less favorable for wheat cultivation. Much less effort has been invested in attempts to improve the milling and baking properties and, as a result, the requirements for foods for humans have not been closely approached. This has caused frustration among some of those involved in the quest to make triticale an acceptable human food. However, slow progress in science needs to be understood in relation to the difficulty of the problem. It illustrates the need for a systematic, multidisciplinary attack.

Aspects of functionality are polygenic traits. Therefore, some of the newer genetic approaches, such as transformation, have limitations. If the composition required to produce certain functional properties is not well understood, genetic engineers are working in the dark to manipulate the genetic composition. Before a viable plan to adjust the genetic makeup of a cereal variety is devised, it is essential to know how the functional properties will be affected by altering the many gene products.

The most direct approach to determining composition–functionality relationships is by application of cereal chemistry. This involves accurately measuring the composition of the grain and flour. Then, a comparison needs to be made with the composition of wheat grain/flour because this is the yardstick for achieving the functionality required to make products similar to the many highly accepted products from wheat. The types of strategies outlined in Chapter 12 are required to achieve the desired goals. Development of a nutritious, high-yielding cereal species with good processing quality is a worthy goal for cereal scientists. It requires great patience, ingenuity, and good collaboration between the different disciplines. The contribution of the cereal chemist to the task should not be underestimated.

References

Bacic, A., B. Stone, and G. B. Fincher. 2009. *Chemistry, biochemistry and biology of 1-3 beta-glucans and related polysaccharides.* New York: Academic Press.

Baik, B.-K., and S. Ulrich. 2008. Barley for food: Characteristics, improvement and renewed interest. *Journal of Cereal Science* 48:232–242.

Bao, J., M. Sun, D.-X. Zhu, and H. Corke. 2004. Analysis of quantitative trait loci for some starch properties of rice (*Oryza sativa* L.): Thermal properties, gel texture and swelling volume. *Journal of Cereal Science* 39:379–385.

Bondia-Pons, I., A. M. Aura, S. Vuorella, M. Kolehmainen, H. Mykkanen, and K. Poutanen. 2009. Rye phenolics in nutrition and health. *Journal of Cereal Science* 49:323–336.

Brennan, C. S., N. Haris, D. Smith, and P. R. Shewry. 1996. Structural differences in the mature endosperm of good and poor malting barley cultivars. *Journal of Cereal Science* 24:171–177.

Bushuk, W., and F. MacRitchie. 1989. Wheat proteins: Aspects of structure that determine bread-making quality. In *Protein quality and the effects of processing*, ed. R. Dixon Phillips and J. W. Finley, 345–369. New York: Marcel Dekker.

Chernyshova, A. A., P. J. White, M. P. Scott, and J.-L. Jannink. 2007. Selection for nutritional function and agronomic performance in oat. *Crop Science* 47:2330–2339.

Darlington, H. F., L. Tecsi, N. Harris, D. L. Griggs, I. C. Cantwell, and P. R. Shewry. 2000. Starch granule associated proteins in barley and wheat. *Journal of Cereal Science* 32:21–29.

Edney, M. J., T. M. Choo, D. Kong, T. Ferguson, K. M. Ho, K. W. May, and R. A. Martin. 1998. Kernel color varies with cultivars and environments in barley. *Canadian Journal of Plant Science* 78:217–222.

Ferrari, B., F. Finochiaro, A. M. Stanca, and A. Giannetti. 2009. Optimization of air classification for the production of β-glucan-enriched barley flours. *Journal of Cereal Science* 50:152–158.

Flander, L., M. Salmenkalio-Martilla, T. Suorrti, and K. Autio. 2007. Optimization of ingredients and baking process for improved oat bread quality. *Food Science and Technology* 40:860–870.

Gautier, M. F., P. Cosson, A. Guirao, R. Alary, and P. Joudrier. 2000. Puroindoline genes are highly conserved in diploid ancestor species but absent in tetraploid *Triticum* species. *Plant Science* 153:81–91.

Glidewell, S. M. 2006. NMR imaging of developing barley grains. *Journal of Cereal Science* 43:70–78.

Gray, D. A., R. H. Auerbach, S. Hill, R. Wang, G. M. Campbell, C. Webb, and J. B. South. 2000. Enrichment of oat antioxidant activity by dry milling and sieving. *Journal of Cereal Science* 32:89–98.

Gupta, H. S., P. K. Agrawal, V. Mahajan, G. S. Bisht, A. Kumar, P. Verma, A. Srivastava, R. Babu, M. C. Pant, and V. P. Mani. 2009. Quality protein maize for nutritional security: Rapid development of short duration hybrids through molecular marker assisted breeding. *Current Science* 96:230–237.

Hamaker, B. R., and B. A. Bugusu. 2003. Overview: Sorghum proteins and food quality. In *Afripro workshop on the proteins of sorghum and millets: Enhancing nutritional and functional properties for Africa*, ed. P. S. Belton and J. R. N. Taylor. Pretoria, South Africa, April 2–4, 2003.

Hamaker, B. R., A. A. Mohamed, J. E. Habben, C. P. Huang, and B. A. Larkins. 1995. Efficient procedures for extracting maize and sorghum kernel proteins reveals higher prolamin contents than the conventional method. *Cereal Chemistry* 72:583–588.

Heinio, R. L., K.-H. Liukkonen, O. Myllymaki, J. M. Pihlava, H. Adlercreutz, S.-M. Heinonen, and K. Poutanen. 2008. Quantities of phenolic compounds and their impacts on the perceived flavor attributes of rye grain. *Journal of Cereal Science* 47:566–575.

Horigane, A. K., N. M. H. G. Engelaar, S. Maruyama, M. Yoshida, A. Okubo, and T. Nagatu. 2001. Visualization of moisture distribution during development of rice caryopses (*Oryza sativa* L.) by nuclear magnetic resonance microimaging. *Journal of Cereal Science* 33:105–114.

Ishimaru, T., A. K. Horigane, H. Ida, A. Iwasawa, Y. A. San-Oh, M. Nakazono, N. K. Nishizawa, T. Masumura, M. Kondo, and M. Yoshida. 2009. Formation of grain chalkiness and changes in water distribution in developing rice caryopses grown under high-temperature stress. *Journal of Cereal Science* 50:166–174.

Izquierdo, L., and I. D. Godwin. 2005. Molecular characterization of a novel methionine-rich-δ-kafirin seed storage protein gene in sorghum (*Sorghum bicolor* L.). *Cereal Chemistry* 82:706–710.

Jiang, H., W. Diane, and P. Wu. 2003. Effect of high temperature on fine structure of amylopectin in endosperm by reducing the activity of starch branching enzyme. *Photochemistry* 63:53–59.

Kim, S. S., S. E. Lee, O. W. Kim, and D. C. Kim. 2000. Physicochemical characteristics of chalky kernels and their effects on sensory quality of cooked rice. *Cereal Chemistry* 77:376–379.

Lawton, J. W. 1992. Viscoelasticity of zein–starch doughs. *Cereal Chemistry* 69:351–355.

Lazstity, R. 1996. *The chemistry of cereal proteins,* 2nd ed., 185–225. Boca Raton, FL: CRC Press.

Li, J., M. Baga, B. G. Rosenagel, W. G. Legge, and R. N. Chibbar. 2008. Identification of quantitative trait loci for β-glucan concentration in barley grain. *Journal of Cereal Science* 48:647–655.

Liang, J., Z. Li, K. Tsuji, K. Nokano, M. J. R. Nout, and R. J. Hamer. 2008. Milling characteristics and distribution of phytic acid and zinc in long-, medium-, and short-grain rice. *Journal of Cereal Science* 48:83–91.

Lin, S. K., M. C. Chang, Y. G. Tsai, and H. S. Leir. 2005. Proteomic analysis of the expression of proteins related to rice quality during caryopsis development and the effect of high temperature expression. *Proteomics* 5:2140–2156.

Lou, J., L. Chen, G. Yue, Q. Lue, H. Mei, L. Xiong, and L. Lue. 2009. QTL mapping of grain quality traits in rice. *Journal of Cereal Science* 50:145–151.

Ong, M. H., and J. H. V. Blanshard. 1995. Texture determinants in cooked parboiled rice. I. Rice starch amylose and the fine structure of amylopectin. *Journal of Cereal Science* 21:251–260.

Perdon, A. A., T. J. Siebenmorgen, R. W. Buescher, and E. E. Gliurr. 1999. Starch retrogradation and texture of cooked milled rice during storage. *Journal of Food Science* 64:828–831.

Peterson, D. M. 2001. Oat antioxidants. *Journal of Cereal Science* 33:115–129.

Prasanna, B. M., S. K. Vasal, K. B. Kasahun, and N. N. Singh. 2001. Quality protein maize. *Current Science* 81:1308–1319.

Sato, K., and K. Inaba. 1976. High-temperature injury of ripening in rice plant. V. On the early decline of assimilate storing ability of grains at high temperature. *Proceedings of the Crop Science Society of Japan* 45:156–161.

Shewry, P. R. 2007. Improving the protein content and composition of cereal grain. *Journal of Cereal Science* 46:239–250.

Shewry, P. R., A. S. Tatham, J. Forde, M. Kreis, and B. J. Miflin. 1986. The classification and nomenclature of wheat gluten proteins: A reassessment. *Journal of Cereal Science* 4:97–106.

Singh, N., L. Kaur, S. Sodhi, and K. S. Sekhon. 2005. Physicochemical, cooking and textural properties of milled rice from different Indian rice cultivars. *Food Chemistry* 89:253–259.

Taylor, J. R., T. J. Schober, and S. R. Bean. 2006. Novel food and nonfood uses for sorghum and millets. *Journal of Cereal Science* 44:252–271.

Vikki, L., L. Johansson, M. Yiven, S. Mauno, and P. Ekholm. 2004. Structural characterization of oats and barley. *Carbohydrate Polymers* 59:357–366.

Wong, J. H., T. Lau, N. Cai, J. Singh, J. F. Pederson, W. H. Vensel, W. J. Hurkman, J. D. Wilson, P. G. Lemaux, and B. B. Buchanan. 2009. Digestibility of protein and starch from sorghum (*Sorghum* bicolor) is linked to biochemical and structural features of grain endosperm. *Journal of Cereal Science* 49:73–82.

Yamakawa, H., T. Hirose, M. Kuroda, and T. Yamaguchi. 2007. Comprehensive expression of profiling of rice grain ripening-related genes under high temperature using DNA microarray. *Plant Physics* 144:258–277.

Yu, S., Y. Ma, and D.-W Sun. 2009. Impact of amylose content on starch retrogradation and texture of cooked milled rice during storage. *Journal of Cereal Science* 50:139–144.

Zegada-Lizarazu, W., and M. Iijima. 2005. Deep root penetration may help millet species to exploit soil water more efficiently and therefore overcome drought stress. *Plant Production Science* 8:454–460.

chapter fourteen

Health aspects of cereals

Whole grains

Research has demonstrated the health benefits of whole grains in the human diet. This research is difficult to carry out because it depends on monitoring the health of human subjects over long periods as well as minimization of the many variables that can interfere with the conclusions. Increase of whole grain in the diet has been found to be associated with reduction in common diseases such as cardiovascular disease, type 2 diabetes, and some forms of cancer. Some of the beneficial chemical components have been identified. However, some evidence indicates that addition of these components to fortify grain products may not always be as effective as having the components in their original state. This has led to a belief that the observed health benefits may not be due as much to specific components as to their action in combination with other components (i.e., the total package).

Refining of cereal grains usually involves removal of bran and germ to leave the endosperm. There are good reasons for doing this. Germ contains high concentrations of lipids, which can have negative effects on shelf life due to their susceptibility to rancidity. Presence of bran in refined flour may detract from the color and palatability of products made from it. As we have seen in Chapter 3, the fruit coat and seed coat of wheat comprise many different layers with varying composition, and some of these (e.g., the outer pericarp) may have detrimental effects in bakery products.

However, the concentrations of minerals, vitamins, and phytochemicals are very high in the bran and their removal lowers the nutritional value. Some data for the concentrations of vitamins and minerals in the starchy endosperm and the aleurone components are shown in Table 14.1. Myriad different chemical compounds can be found in the whole grain of cereals, some of which may not yet have been identified. Therefore, cereal chemists have a great challenge for the future to identify them and to contribute to elucidating their role in improving health.

Health-related components of cereal grains

There is a vast literature on cereal components that have been shown to give health benefits. The main ones will be briefly summarized.

Table 14.1 Comparison of Some Vitamins in
Endosperms and Aleurone Tissue of Wheat

	Endosperm	Aleurone
Vitamin		
Niacin	8.5 mg/g	614 mg/g
Thiamine	0.07 mg/g	36 mg/g
Pantothenic acid	0.7 mg/g	45 mg/g
Raboflavin	39.0 mg/g	10 mg/g
Mineral		
Manganese	1.5 mg/g	78 mg/g
Phosphorus	0.054%	3.17%
Iron	11 mg/g	338 mg/g
Calcium	0.012%	0.404%
Copper	7 mg/g	51.5 mg/g

Dietary fiber

One of the problems highlighted in recent times is the importance of fiber in the diet. Normal starch is rapidly broken down into simple sugars by enzymes (amylases) in the human body. A high consumption of starch can lead to spikes in blood sugar level. The rate and extent of this rise are measured by the glycemic index. To avoid sharp fluctuations, which are thought to be damaging to health, inclusion of a certain amount of fiber in the diet is recommended.

Insoluble fibers include the polysaccharides cellulose and hemicellulose. They facilitate the movement of waste through the digestive tract. Examples of soluble fibers are the β-glucans—also polysaccharides—that have attracted a good deal of attention due to the claim that they lower blood cholesterol levels. Beta-glucans are found in most cereals and occur at relatively high levels in the bran of oats and barley (See Chapter 13).

Resistant starch has also generated interest in recent times. This is starch that escapes digestion in the small intestine of healthy individuals and passes through the large intestine, where it acts like dietary fiber. Resistant starches have been classified into four main groups:

1. physically inaccessible starch, as found in unprocessed whole grain
2. starch that occurs in its natural granular form, such as in high amylose corn; amylose is more resistant to digestive enzymes than amylopectin
3. starch from foods that are cooked and cooled, such as bread or cornflakes; on cooling, the starch retrogrades (i.e., becomes more crystalline)
4. starch that has been chemically modified (e.g., cross-linked starch)

Vitamins and minerals

The concentrations of vitamins and minerals are very much greater in the bran than the endosperm of cereals (cf. Table 14.1). As a result, bran is being utilized to manufacture neutroceuticals (nutritional supplements). A serious problem is the deficiency of some minerals, particularly in developing countries. Iron is one mineral that is widely deficient; lack of iron leads to problems associated with anemia.

Phytic acid (or phytate in the salt form) is present in cereal bran layers and can have beneficial health benefits as an antioxidant. However, it can also act as an antinutrient by binding important minerals such as calcium, magnesium, iron, and zinc. The main antinutritional effect of phytate is that it makes phytate phosphorus unavailable for digestion and absorption by nonruminants. Phytase is an enzyme that breaks down phytate. Thus, phytate and phytase are two major factors associated with the bioavailability of iron and zinc in cereal-based foods. An interesting approach to determining the distribution of phytate, phytase activity, iron, and zinc in different bran layers of wheat has been reported by Liu et al. (2006), who used pearling to separate the layers and analyze their compositions.

Antioxidants

Free radicals may be essential to metabolic processes, but they can also damage cellular material at the molecular level. If oxidized, protein, lipid, lipid membranes, and DNA may initiate degenerative diseases. Because they have one or more unpaired electrons, free radicals are highly unstable. They scavenge the body to grab or donate electrons, thereby damaging cells. Antioxidants can counteract the damaging effects by blocking the process of oxidation and neutralizing free radicals; in doing so, they become oxidized. Some natural antioxidants, such as vitamin E and beta-carotene, have been extensively studied.

Grains contain a wide range of antioxidants, including phenolic acids, flavonoids, and saponins. Phenolic acids are common antioxidants in grain. They exist primarily as benzoic and cinnamic acid compounds. Flavonoids are more concentrated in fruits and vegetables but are also found in grains. They number in the thousands and have a basic three-ring structure; however, activity varies greatly depending on the number and location of hydroxyl groups. Typically, phenolic acids and flavonoids are water-soluble compounds; lipid-soluble derivatives are common in grains such as caffeic and ferulic acid esters of C_{20}–C_{28} chain mono- and dialcohols.

Several methods are used to measure antioxidant content of extracts. A simple method is to use 2,2-diphenyl-1-picrylhydrazyl, a stable free radical, as the detection agent. In this work, extraction efficiency is crucial. Grain antioxidants vary in solubility from water soluble to lipid soluble and many are covalently bound to cell wall material. Bran and germ, which contain the highest concentrations of antioxidants, have a thick-walled cellular structure that inhibits solvent extraction.

Antinutritional factors

Not all constituents of cereals are advantageous for health. Although many nutritionally beneficial components are concentrated in the outer layers of the grain, it may also happen that residues from crop sprays such as pesticides or fumigants from stored grain may remain. This may pose a problem for processors, who must also ensure that these residues do not become concentrated in the products. Other hazards include mycotoxins, which can be dangerous at low levels.

Skills in separation of grain components and their analysis are needed to address these problems. Some examples of naturally occurring antinutritional compounds are phytates (already mentioned), phenols, tannins, and trypsin- and α-amylase inhibitors. Phenols and tannins bind with and precipitate proteins in food systems, thus decreasing digestibility. Minimization of these compounds in cereals is an objective.

Other antinutritional compounds may form during processing. One that has attracted attention is acrylamide, which is believed possibly to be a neurotoxic carcinogen. Acrylamide has been detected in starch-enriched foods heated to high temperatures (Slayne and Lineback 2005). Other unhealthy constituents may be exacerbated in cereals by environmental influences. For example, high concentrations of arsenic (observed in rice especially) are creating problems in different parts of the world. The origin of this problem is often traced to irrigation water containing high levels of arsenic from the soil or from wells.

Metabolism chemistry

The chemical reactions that participate in metabolism are numerous and coordinated. Cereal chemists have a role in identifying the constituents provided by cereals in the diet and how they may interact. Basically, there are two main types of metabolic processes:

1. Catabolic processes involve breaking down of complex compounds in the diet. These compounds can then pass through the lining of the digestive system into the blood, where they are carried to the cells. Thus, proteins, carbohydrates, and fats are degraded by proteases, amylases, and lipases into amino acids, simple sugars, and glycerol and fatty acids, respectively.
2. After these substances are transported to the cells, *anabolic* reactions, the second main type of reaction, occur. In this phase of metabolism,

energy is used to build up the complex molecules needed to maintain and develop the living system.

Unraveling of the reaction pathways that influence healthy and nonhealthy effects of dietary compounds promises to be an exciting and challenging area of research for the future.

Gluten intolerance and the question of viscoelastic properties for nonwheat cereals

The problem of gluten intolerance is increasingly recognized and is regarded as having been underdiagnosed. Although it was believed to affect only a small segment of the population, it is becoming clear that the problem is more widespread than previously thought (Braly and Hoggan 2002). Celiac disease is the best known ailment associated with intake of gluten. It is caused by specific sequences of amino acids in gluten proteins (Kasarda 2000). The only cure is strictly avoiding consumption of gluten products in the diet. Thus, wheat, rye, and barley need to be excluded. Other cereals such as rice, maize, sorghum, millet, and, possibly, oats are free of this health problem.

This problem has led to the development of food products from other cereals to replace those containing gluten. However, these products do not match the palatability of aerated products such as the different wheat breads prized by consumers. One problem is that the nonwheat cereals do not give dough with the viscoelastic properties required to incorporate and stabilize a finely dispersed gas cell structure. If this problem could be overcome for some of the nonwheat cereals, it could lead to attractive aerated products for the gluten-intolerant section of the population. It might also lead to increased utilization of these cereals for human food, particularly because some (e.g., sorghum) can be grown in areas not suitable for wheat. We will therefore now consider the requirements for introducing viscoelastic dough properties into cereals.

Requirements for viscoelastic properties

The reason why wheat is the only cereal that gives dough that exhibits viscoelastic properties ideal for production of high-volume aerated products is a fundamental question that has challenged cereal chemists for a long time. It is one of the questions posed in Chapter 1. Much has been learned from polymer science about the requirements for a material to acquire viscoelastic properties. This knowledge can be applied to investigate possible approaches for introducing viscoelasticity in cereals.

A polymer material will show viscoelastic properties if the ambient temperature is above its glass transition temperature (T_g). The relationship between compliance (inverse of elastic modulus) and time for a polymer undergoing creep was illustrated in Figure 6.2 (Chapter 6). At low temperatures, the polymer exists as a glass. As the temperature is raised, the T_g is exceeded, the compliance increases, and the polymer takes on leathery and rubbery properties. The value of T_g is influenced by the presence of plasticizers—compounds of low MW that reduce the elastic modulus. For dough, water is the most relevant plasticizer.

The importance of T_g in food systems has been emphasized in the work of Levine and Slade (1989). Figure 6.3 in Chapter 6 is another generalized diagram useful for illustrating how a cereal protein might behave as a function of water content and temperature. Here, T_g is the glass transition temperature at no moisture and T_g' is that at which free (freezable) water starts to appear as a separate phase (W_g'). If we begin at the point A, the vertical arrow B shows the effect of increasing the temperature and the horizontal arrow shows the effect of increasing the water content. The rubbery region where viscoelastic properties appear can thus be reached by increasing the temperature at constant water or by increasing the water content at constant temperature.

The continuous gluten protein network in wheat flour dough contributes the viscoelastic properties. The T_g of dry gluten is high—above 120°C (Hoseney, Zeleznak, and Lai 1986)—and decreases steeply as the water content is increased. It reaches a value close to ambient temperature at a water content of about 16%. The water content of gluten in a dough is about 45% (i.e., at this water content, the protein is well above its T_g). The T_g–temperature relationship for the maize protein zein is similar to that of gluten but, for given water contents, the T_g is higher (Lawton 1992). When heated above 30°C, zein–starch dough exhibits viscoelastic properties (Lawton 2002; Carson 2005). Heating above ambient temperature is required to surpass the T_g of zein. On cooling, the dough loses its viscoelasticity as its temperature decreases below the T_g.

Requirements for aerated products for cereals

In order for a cereal to give high-volume aerated products, its protein must contribute to viscoelastic dough properties. This enables the incorporation of the gas bubbles during mixing that allows the dough to expand. This is the case for dough of a zein–starch mixture, described in the previous paragraph. However, when this dough is fermented and baked, there is little expansion of the dough, as is desired in a high-volume baked product. Thus, dough viscoelasticity must not be the only criterion to be met.

Molecular weight distribution

The balance between viscosity and elasticity is close to optimum for wheat gluten protein at the water content used in dough-making. We have seen that the first requirement for a protein in dough to have viscoelastic properties is that its T_g be below the processing (usually ambient) temperature for the water content used. A second requirement for dough in an aerated product is that the molecular weight distribution (MWD) of the protein be optimal.

As we have seen in Chapter 7, it appears that a high strain hardening index and a high failure strain are essential for dough to have good expansion capacity. If the protein MWD is shifted too far toward lower values, the dough will be weak and lack the high strain hardening properties required. Similarly, if the MWD is shifted to values that are too high, the dough will lack extensibility and a high strain hardening index will not be attained.

Zein is a prolamin analogous to the gliadin of wheat. A zein–starch dough thus lacks the glutelin component of wheat (glutenin) that would be needed to provide the large molecules that contribute to strain hardening. Lack of a suitable balance of proteins may thus occur as a result of a nonoptimum ratio of prolamins to glutelins. Disruption of the network may also occur by the presence of high concentrations of nonfunctional proteins (e.g., globulins) or fiber components (e.g., pentosans).

Liquid lamellae stability

A third requirement may be related to the secondary stabilizing mechanism discussed in Chapter 7. Wheat flour contains a relatively high proportion of polar lipids. These lipids, comprising galactolipids and phospholipids, combine with protein to give a good stabilizing film at the gas–liquid interface of the gas bubbles. Some other cereals have lower amounts of polar lipids. The nonpolar lipids, such as mono-, di-, and tri-glycerides and free fatty acids (in particular), tend to act as antifoaming agents. A high concentration of these lipids will therefore act to destabilize the gas cell structure. This is a third factor that needs to be considered when attempting to utilize nonwheat cereals for aerated products.

References

Braly, J., and R. Hoggan. 2002. *Dangerous grains* (Avery Publishing Group). *Lancet* 2000 356:399–400.

Carson, B. 2005. Viscoelastic properties of zein. M.S. thesis, Kansas State University, Manhattan, KS.

Hoseney, R. C., K. Zeleznak, and C. S. Lai. 1986. Wheat gluten: A glassy polymer. *Cereal Chemistry* 63:285–286.

Kasarda, D. D. 2000. Celiac disease. In *The Cambridge world history of food,* vol. 1, ed. K. F. Kiple and C. Ormelas, 1008–1022. Cambridge, England: Cambridge University Press.

Lawton, J. W. 1992. Viscoelasticity of zein–starch doughs. *Cereal Chemistry* 69:351–354.

———. 2002. Zein: A history of processing and use. *Cereal Chemistry* 79:1–18.

Levine, H., and L. Slade. 1989. Influences of the glassy and rubbery states on the thermal, mechanical, and structural properties of doughs and baked products. In *Dough rheology and baked product texture,* ed. H. Faridi and J. M. Faubion, 157–330. New York: Van Nostrand Reinhold.

Liu, Z. H., H. Y. Wang, X. E. Wang, G. P. Zhang, P. D. Chen, and D. J. Liu. 2006. Phytase activity, phytate, iron, and zinc contents in wheat pearling fractions and their variation across production locations. *Journal of Cereal Science* 45:319–326.

Slayne, M. A., and D. R. Lineback. 2005. Acrylamide: Considerations for risk management. Acrylamide in food: A survey of two years of research activities. *Journal of AOAC International* 88:227–233.

chapter fifteen

The scientific method

Many active scientists spend little time thinking about the philosophy of science and, in science courses at universities, the subject is rarely, if ever, covered. Students often embark on a scientific research project without having devoted much time to trying to understand what is involved in the scientific method. For that reason, this short chapter will provide a number of questions that are intended to stimulate discussion and encourage thought about the topic. The questions do not necessarily have correct or unequivocal answers but, in many cases, should lead to healthy debate. For that reason, no answers will be suggested.

1. What distinguishes the scientific method from other methods of enquiry?
2. Is science empirical (i.e., based on observations)?
3. Is science inductive (i.e., makes a series of observations and then attempts to arrive at a generalization from these observations)?
4. What are some of the characteristics of the scientific method?
5. Does science proceed from observation to theory or from theory to observation?
6. Can a scientific hypothesis or theory be proved?
7. Can a scientific hypothesis or theory be disproved (refuted)?
8. Which is more important in experimental measurements: accuracy or precision?
9. If a few measurements do not fit with a theory, should these points be omitted?
10. If experimental measurements do not support a theory, should the measurements be deleted or should the theory be rejected? Conversely, should the theory be modified to fit the measurements?
11. What makes a good scientific hypothesis?
12. Does progress in science mean a continuous accumulation of information OR refined and new theories that increase understanding?
13. What are some of the attributes required for a successful research scientist?

Bibliography

MacRitchie, F. 1994. Scientific research: Philosophy and organization. *Chemistry in Australia* 61(1):14–15.

Popper, K. R. 1962. Science: Conjectures and refutations. In *Conjectures and refutations: The growth of scientific knowledge,* 33–65. New York: Basic Books.

Suggested solutions to exercises

Chapter 4

1. Most of the standard methods for measuring grain hardness use several grams of grain. Estimation of hardness requires some ingenuity when only a few kernels are available. Some instruments, such as a TA XT2 or an Ogawa Seiki O.S.K. grain hardness tester (Kiya Seisakusho Ltd., Tokyo), can measure the maximum force when a single kernel is compressed. The standard deviation of these measurements is high, but an estimate of relative hardness for different samples might be obtained. The degree of swelling in different solvents has been related to hardness of polymers, which may be an indirect approach.

2. The affinity of proteins for a polar–nonpolar interface is governed by their hydrophobic/hydrophilic balance. In an immature wheat kernel, the aqueous phase is polar and the starch granule surface, although it perhaps might not be classified as nonpolar, is probably more nonpolar than water. Thus, the more nonpolar that a protein is, the greater its distribution will favor the starch granule surface. Hydrophobicities have been calculated for amino acid residues. The values may change according to the method of calculation but the pattern seems clear. Based on one hydrophobicity table in the literature, the changes as a result of the three mutations are as follows:

glycine to serine: −0.4 to −0.8; leucine to proline: 3.8 to −1.6; tryptophan to arginine: −0.9 to −4.5. The higher the value (i.e., the more
positive), the greater is the hydrophobicity. Therefore, for each of the
amino acid residue changes, the hydrophobicity decreases or the
polarity increases. This is consistent with a decrease in concentration of puroindoline at the starch granule surface.

Chapter 6

1. The requirement for viscous flow is that molecules move relative to
   one another in response to a stress. This is what happens in a liquid
   such as water. Vulcanized rubber is a polymer in which molecules are
   cross-linked. Chain segments between cross-links can be stretched.
   However, the cross-links prevent net flow. Similar behavior occurs
   when the gluten matrix in a dough is stretched. The essential difference with rubber is that the gluten is not cross-linked, but rather has
   large molecules that are entangled. Entanglements act as transient
   cross-links. If a stress is applied, there is a restoring elastic force due
   to extension of chain segments; however, unlike cross-linked systems, chains can slip through entanglements so that there can be a
   net relative movement of molecules.

Chapter 7

1. The answer to this question requires researching the literature to find
   whether the behavior of these fatty acids at the gas–liquid interface is dif-
   ferent. A good reference to help with understanding behavior of mono-
   layers is Gaines, G. L. 1966. *Insoluble Monolayers at Liquid–Gas Interfaces.*
   Ann Arbor, MI: Interscience Publishers, University of Michigan.
2. Area/molecule = total area/total number of molecules
   Total area = 162.8×10^{16} Å^2
   Total number of molecules = 0.036 mL × 0.001 g/mL × 6.02×10^{23}/MW
   Therefore, 20.5 = $(162.8 \times MW)/0.036 \times 0.001 \times 6.02 \times 10^{23}$
   Therefore, MW = 272.9
   This value is close to the molecular weight (MW) of octadecanol (270).

Chapter 8

1. Some observations are as follows:
   Bread crumb begins to firm after removal from the oven.

Bread firms in the absence of moisture loss.

Rate of crumb firming is greatest near the center of the loaf.

Bread can be refreshened by heating above the starch gelatinization temperature.

Firmness increases after several cycles of heating and cooling.

Crumb firming can be practically eliminated by storage at freezer temperature (–20°C).

Crumb firming is inversely proportional to specific volume of bread.

Rate of crumb firming is at a maximum near (but not at) the freezing point of water and decreases as temperature is increased or decreased on either side.

Crumb firming can be reduced by additives such as enzymes and surfactants.

The higher the flour protein content is, the lower the rate of crumb firming will be.

2. Starch granules have a high concentration in dough and bread. Gelatinization causes swelling, evidently causing some fusion of granules so that they can no longer be considered completely dispersed.

3. This exercise may not have practical relevance, but it is good training for tackling problems. In order to arrive at an estimate, it is necessary to make some assumptions. For example, assume that (a) gas bubbles are spherical and a certain size, (b) temperature in the loaf is 100°C, and (c) activity of water is unity. The steps would then be

Calculate average radius of gas bubbles in bread.

Find concentration of water vapor in bubble at baking temperature (say, 100°C) from properties of saturated steam in chemical or physical handbook.

Find concentration of water that condenses. This will be equal to concentration of water vapor at 100°C – concentration at 20°C.

Find volume of water that condenses in bubble.

Find surface area of bubble.

Calculate thickness of water layer from volume of water condensed/area of bubble surface.

Chapter 9

1. One obvious source of entropy increase is the release of water of crystallization from $Ba(OH)_2$. A large increase in disorder will result from the freeing of water molecules, thus contributing to a greater entropy of mixing.

2. Shaking can cause fluctuations in the degree of mixing. However, as a fluctuation moves the system farther from the most disordered state, the probability of its occurring becomes lower. It is more correct to say that it is not impossible for the original ordered state to be recovered, but that the probability of its occurring by random rearrangement becomes infinitesimally low.

3. (a) ΔH contributes to a lowering of ΔG. If the protein were soluble, ΔG would also be lowered as a result of the entropy of mixing. Because the protein is not soluble, there must be another contribution due to an entropy decrease that outweighs the contributions of the negative ΔH and the entropy of mixing to make ΔG positive. This could be due to an increase in ordering of the protein or the water molecules (as in the hydrophobic effect).

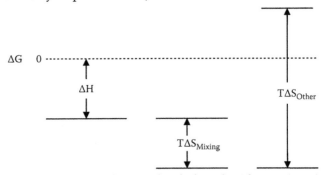

(b) The only change that can be deduced with certainty is that the entropy of mixing must decrease due to the lesser number of protein molecules and therefore the lesser number of possible arrangements of protein and solvent molecules. The entropy of mixing must, however, always contribute to a negative free energy even though it is decreased. Therefore, there must be a contribution of greater magnitude due either to a positive ΔH or some other entropy change (or both).

(c) The net electrical charge on a protein is zero at its isoelectric point and increases as the pH is moved to higher or lower values. As the net charge increases, the strength of interaction with water molecules increases, thus lowering the free energy of the system. This may be due to a decrease in enthalpy, but entropic changes may also occur.

Chapter 11

1. Mixograph dough development time (MDDT) may be used as a measure of dough strength. The decrease in MDDT with deletion of high molecular weight glutenin subunits (HMW-GS) can be estimated from the slope of MDDT versus n (number of HMW-GS) from the data in Table 1 of Lawrence, MacRitchie, and Wrigley 1988. *Journal of Cereal Science* 7:109–112. The change in MDDT as a result of allelic variation in near-isogenic lines (HMW-GS 5+10 to 2+12) can be calculated from the results in Figure 2 from Gupta and MacRitchie 1994. *Journal of Cereal Science* 19:19–29.

2. Deletion of the *Gli-B1/Glu-B3* locus of the three parent lines results in a decrease of the percentage of polymeric protein (PPP). It can therefore be deduced that the amount of low molecular weight glutenin subunits (LMW-GS) is greater than gliadins coded by this locus because its deletion reduces the glutenin/gliadin ratio. It should be noted that this may not be a general result. Deletions of *Gli-1/Glu-3* loci on the A and D chromosomes or even deletions of *Gli-1/Glu-3* on other cultivars could conceivably have the opposite effect. This is important to know if deletions of these loci are used in breeding programs to shift the glutenin/gliadin ratio in a certain direction.

Chapter 12

1. The HMW/LMW-GS ratio is relatively high (Gupta, Batey, and MacRitchie 1992. *Cereal Chemistry* 69:125–131); however, due mainly to a dearth of LMW-GS (Singh, Donovan, and MacRitchie 1990. *Cereal Chemistry* 67:161–170), its glutenin/gliadin ratio is unusually low. The high HMW/LMW-GS ratio contributes to high molecular weights and therefore relatively high R_{max}. The low ratio of glutenin/gliadin may be responsible for the sticky dough properties.

Index

.

Printed and bound by CPI Group (UK) Ltd, Croydon, CR0 4YY

23/10/2024

01777696-0019